ORDINARY DIFFERENTIAL EQUATIONS AND APPLICATIONS
Mathematical Methods for Applied Mathematicians, Physicists, Engineers, Bioscientists

DEDICATION

In memory of Luise Podolan and Luise Meier (WSW)

and

To Grace, Alan and Fraser (KAL)

Mathematics possesses not only truth, but supreme beauty—a beauty cold and austere, like that of sculpture, and capable of stern perfection, such as only great art can show.

Bertrand Russell in *the Principles of Mathematics* [1872-1970]

Talking of education, "People have now a-days" (said he) "got a strange opinion that every thing should be taught by lectures. Now, I cannot see that lectures can do so much good as reading the books from which the lectures are taken. I know nothing that can be best taught by lectures, expect where experiments are to be shewn. You may teach chymestry by lectures. — You might teach making of shoes by lectures!"

James Boswell: *Life of Samuel Johnson, 1766* [1709-1784]

ABOUT OUR AUTHORS

Werner S. Weigelhofer obtained the degrees of Dipl.-Ing and Dr. techn. from the Technical University of Graz, Austria in 1985 and 1986 respectively. After a year of postdoctoral research at the University of Adelaide, Australia, he joined the University of Glasgow in 1988 as a Research Assistant in the Department of Mathematics, where he is currently a Senior Lecturer. In his research area, electromagnetic field theory, he has authored or co-authored some 100 refereed journal publications and is currently writing a monograph on electromagnetic fields in materials.

Kenneth A.Lindsay obtained the degrees of B.Sc. from the University of Glasgow, Scotland, in 1970 and D.Phil. from Merton College at the University of Oxford in 1973. He joined the University of Glasgow in 1973 as a Lecturer in the Department of Mathematics, where he was promoted to his current appointment as a Reader. From his research base in continuum mechanics, he has contributed to elasticity, non-linear wave propagation and convection in fluids. His current interests lie in the application of stochastic differential equations to economics, oncology and neurophysiology.

Ordinary Differential Equations and Applications

Mathematical Methods for Applied Mathematicians, Physicists, Engineers, Bioscientists

Werner S. Weiglhofer Dipl.-Ing., Dr.techn

and

Kenneth A. Lindsay B.Sc., D.Phil
Department of Mathematics
University of Glasgow

WOODHEAD
PUBLISHING

Oxford   Cambridge   Philadelphia   New Delhi

1

Differential Equations of First Order

1.1 GENERAL INTRODUCTION

A general differential equation of first order is an expression of the type

$$F(x, y, y') = 0.\qquad (1.1)$$

Therein, x is the **independent** variable, y is the **dependent** variable and $y'(x) = dy/dx$ is the derivative of y with respect to x. Some general functional dependency is symbolized by F permitting *explicit* as well as *implicit* representation of the differential equation. The **general solution** of (1.1), or **general integral**, is

$$f(x, y, C) = 0\qquad (1.2)$$

where C is an arbitrary **integration constant**.

There is a simple geometric interpretation of the aforementioned equations in terms of curves in the $x-y$ plane: (1.1) gives the gradient (explicitly or implicitly) to the curve at every point (x, y) and (1.2) provides an unlimited number of curves that fulfil (1.1).

To single out one **particular solution** of the general solution (1.2), an **initial condition** (so called because the independent variable often signifies time) is needed, say

$$y(x_0) = y_0.\qquad (1.3)$$

The differential equation (1.1) and the initial condition (1.3) define an **initial value problem**. When the initial condition (1.3) is substituted into (1.2), the arbitrary integration constant C can be calculated from

$$f(x_0, y_0, C) = 0.\qquad (1.4)$$

The solution $y(x) \equiv 0$ is called the **trivial solution**.

The goal of the present introductory chapter is to solve (1.1) for various forms of functional dependence F, with and without prescribed initial conditions.

1.2 SEPARABLE EQUATIONS

Equations of the type

$$y' = g(x)h(y) \tag{1.5}$$

are called **separable** (g and h are assumed to be real–valued functions) and have general solution given by[1]

$$\int \frac{dy}{h(y)} = \int g(x)dx. \tag{1.6}$$

To find a particular solution satisfying $y(x_0) = y_0$, it is sufficient to integrate (1.6) and then determine the arbitrary integration constant from the initial condition. Alternatively, the particular solution can be obtained without recourse to the general solution by simply rewriting (1.6) in the definite integral form

$$\int_{y_0}^{y} \frac{dz}{h(z)} = \int_{x_0}^{x} g(t)\,dt \tag{1.7}$$

where z and t are dummy variables.

Example 1.1   Find the general solution of $y' = 3x^2 e^{-y}$ and the particular solution for which $y = 0$ at $x = 0$.

Solution 1.1   From formula (1.6),

$$\int e^y\,dy = \int 3x^2\,dx \qquad \longrightarrow \qquad e^y = x^3 + C$$

and therefore

$$y(x) = \ln(x^3 + C)$$

is the general solution. The initial condition is fulfilled if $0 = \ln(0 + C)$, implying $C = 1$. The particular solution is then

$$y = \ln(x^3 + 1).$$

The direct approach to obtain the particular solution gives

$$\int_{0}^{y} e^z\,dz = \int_{0}^{x} 3t^2\,dt \qquad \longrightarrow \qquad e^y - 1 = x^3$$

which is identical to the result obtained previously.   □

---

[1]We shall use the notations $\int f(x)\,dx$ and $\int^x f(s)\,ds$ interchangeably for indefinite integrals.

1.3   HOMOGENEOUS EQUATIONS

A differential equation

$$P(x,y) + Q(x,y)\,y' = 0 \tag{1.8}$$

is called **homogeneous**[2] if the functions P and Q are homogeneous of degree m, i.e.,

$$P(x,vx) = x^m R(v), \qquad Q(x,vx) = x^m S(v). \tag{1.9}$$

Such an equation is solved by the substitution $y(x) = xv(x)$ where v is a new dependent variable. Since $y' = v + xv'$, then the substitution of $y = xv$ into (1.8), bearing in mind (1.9), gives

$$R(v) + S(v)(v + xv') = 0 \qquad \longrightarrow \qquad v' = -\frac{1}{x}\left[v + \frac{R(v)}{S(v)}\right]. \tag{1.10}$$

This equation is now *separable* and one obtains

$$\int \frac{dv}{[\,v + R(v)/S(v)\,]} = -\ln x + C. \tag{1.11}$$

Example 1.2   Solve $x^2 - y^2 + 2xyy' = 0$.

Solution 1.2   The differential equation is easily recognized to be of homogeneous type with $R(v) = 1 - v^2$ and $S(v) = 2v$. From (1.10), the substitution $y = xv$ now leads to

$$(1 - v^2) + 2v(v + xv') = 0 \longrightarrow 1 + v^2 + 2xvv' = 0$$

$$\longrightarrow \frac{2vv'}{1+v^2} = -\frac{1}{x} \longrightarrow \ln(1 + v^2) = -\ln x + C \longrightarrow (1 + v^2)x = A$$

where A is another arbitrary integration constant.[3] Replacing v by y/x gives the general solution

$$x^2 + y^2 = Ax$$

which can be rewritten in the form

$$(x - B)^2 + y^2 = B^2.$$

Thus the general solution represents circles with centres on the x axis and which touch the y axis at the origin (for $B = 0$ the solution degenerates into the point $x = y = 0$.)   □

---

[2]The reader should be aware that the term *homogeneous* will also be used with a different meaning later in the book.

[3]This is a frequently employed trick to make solutions of differential equations look tidier. If C is an arbitrary constant, any multiple or power or, in general, any function of C is just another arbitrary constant and can be renamed accordingly.

1.4   EXACT EQUATIONS

Let $u(x,y)$ be an arbitrary function of x and y such that

$$u(x,y) = C \qquad (1.12)$$

defines a curve in the $x - y$ plane. The **total derivative** of (1.12) yields

$$\frac{\partial u}{\partial x} + \frac{\partial u}{\partial y}\frac{dy}{dx} = 0 \qquad (1.13)$$

where $\partial u/\partial x$ and $\partial u/\partial y$ are the **partial derivatives** of u with respect to x and y respectively. Now take a differential equation of the form

$$P(x,y) + Q(x,y)\, y' = 0 \qquad (1.14)$$

or, equivalently, $P(x,y)\, dx + Q(x,y)\, dy = 0$. Equation (1.14) is said to be **exact**, if there exists a function $u(x,y)$ such that

$$P(x,y) = \frac{\partial u(x,y)}{\partial x}\,, \qquad Q(x,y) = \frac{\partial u(x,y)}{\partial y}\,, \qquad (1.15)$$

that is, equation (1.14) assumes the form (1.13). It therefore follows from the interchangeability of the order of partial differentiation that a necessary condition for (1.14) to be exact is that

$$\frac{\partial P(x,y)}{\partial y} = \frac{\partial Q(x,y)}{\partial x}\,. \qquad (1.16)$$

Solution recipe. If (1.14) is exact, integrate one of the equations in (1.15), for example the first one with respect to x, and get

$$u(x,y) = \int P(x,y)\, dx + \phi(y) = S(x,y) + \phi(y) \qquad (1.17)$$

where instead of an integration constant, an arbitrary function $\phi(y)$ arises upon integration with respect to x. But the second of equations (1.15) demands that $Q = \partial u/\partial y$. From (1.17) it therefore follows that

$$Q(x,y) = \frac{\partial u(x,y)}{\partial y} = \frac{\partial S(x,y)}{\partial y} + \phi'(y). \qquad (1.18)$$

In the last relation $Q(x,y)$ and $S(x,y)$ are known; this allows calculation of $\phi(y)$ by quadrature, completing the general solution in the form (1.12).
Direct solution. Less obvious, but nevertheless true, is the fact that if P and Q satisfy condition (1.16), then the existence of u in (1.15) is guaranteed. Consider the function

$$u(x,y) = \int_{x_0}^{x} P(z,y)\, dz + \int_{y_0}^{y} Q(x_0,z)\, dz \qquad (1.19)$$

where $y(x_0) = y_0$ is the initial condition for (1.14). We have

$$\frac{\partial u}{\partial x} = P(x, y), \qquad \frac{\partial u}{\partial y} = \int_{x_0}^{x} \frac{\partial P(z, y)}{\partial y} \, dz + Q(x_0, y).$$

However, if P and Q satisfy condition (1.16) then

$$\int_{x_0}^{x} \frac{\partial P(z, y)}{\partial y} \, dz + Q(x_0, y) = \int_{x_0}^{x} \frac{\partial Q(z, y)}{\partial z} \, dz + Q(x_0, y) = Q(x, y).$$

The particular solution of (1.14) is now seen to be $u(x, y) = u(x_0, y_0) = 0$. Thus property (1.16) is a necessary and sufficient condition for exactness of (1.14). Equation (1.19) actually provides an explicit formula for the general solution $u(x, y) = C$ by interpreting x_0 and y_0 as arbitrary.

Remarks 1.1   In the above calculation we have made use of the **fundamental theorem of calculus**, namely

$$\frac{d}{dx} \int_{a}^{x} f(z) dz = f(x).$$ △

Example 1.3   Solve

$$\ln(y^2 + 1) + \frac{2y(x - 1)}{y^2 + 1} y' = 0.$$

Solution 1.3   Partial differentiation leads to

$$\frac{\partial P}{\partial y} = \frac{2y}{y^2 + 1}, \qquad \frac{\partial Q}{\partial x} = \frac{2y}{y^2 + 1}$$

so the criterion for exactness is fulfilled. Then

$$u(x, y) = \int \ln(y^2 + 1) \, dx + \phi(y) = x \ln(y^2 + 1) + \phi(y).$$

Further,

$$\frac{2y(x - 1)}{y^2 + 1} = \frac{2xy}{y^2 + 1} + \phi'(y) \qquad \longrightarrow \qquad \phi'(y) = -\frac{2y}{y^2 + 1}$$

which can be integrated and we obtain

$$\phi(y) = -\ln(y^2 + 1).$$

The general solution is therefore

$$(x - 1) \ln(y^2 + 1) = C.$$

Formula (1.19) gives the same result directly.   □

1.7   BERNOULLI EQUATION

The **Bernoulli** equation is a differential equation of the type

$$y'(x) + p(x)y(x) = q(x)y^n(x) . \qquad (1.25)$$

This equation is linear when $n = 0$ or $n = 1$ but non–linear otherwise. It can be solved by the substitution

$$z = y^{1-n} \qquad \longrightarrow \qquad z' = (1-n)y^{-n}y' \qquad (1.26)$$

for a new dependent variable z. Substitution into (1.25) leads to

$$z'(x) + (1-n)p(x)z(x) = (1-n)q(x) \qquad (1.27)$$

which is now a *linear* equation for $z(x)$ and can be solved accordingly.

Example 1.6   Solve the Bernoulli equation

$$y' - \frac{y}{2x} = 10x^2 y^5 .$$

Solution 1.6   The correct substitution is

$$z = y^{-4} \qquad \longrightarrow \qquad z' = -4y^{-5}y'$$

which transforms the original differential equation into

$$z' + \frac{2}{x}z = -40x^2 .$$

This linear equation has the integrating factor x^2 and thus

$$z(x) = x^{-2} \left[\int^x s^2 \left(-40s^2\right) ds \right] = x^{-2} \left(-8x^5 + C\right) = -8x^3 + \frac{C}{x^2} .$$

Finally,

$$y(x) = \left(\frac{C}{x^2} - 8x^3 \right)^{-1/4} . \qquad \square$$

1.8   RICCATI EQUATION

The equation

$$y'(x) = p(x)y^2(x) + q(x)y(x) + r(x) \qquad (1.28)$$

takes its name from the mathematician **Riccati**. We note also that for $r(x) \equiv 0$, the Riccati equation is a *Bernoulli equation* with $n = 2$. Riccati equations can be integrated if a solution $y(x) = y_1(x)$ is known (and such a solution can often be guessed easily). The substitution

$$y(x) = y_1(x) + \frac{1}{z(x)} \qquad \longrightarrow \qquad y'(x) = y_1'(x) - \frac{z'(x)}{z^2(x)} \qquad (1.29)$$

is then used to introduce a new dependent variable z.

Example 1.7   Solve the Riccati equation $2x^2 y' = (x-1)(y^2 - x^2) + 2xy$.

Solution 1.7   We can see that $y = x$ and $y = -x$ are both solutions of the differential equation. The first choice $y = x$ defines the substitution

$$y = x + \frac{1}{z} \qquad \longrightarrow \qquad y' = 1 - \frac{z'}{z^2}$$

and the original Riccati equation becomes

$$2x^2(z' + z) = 1 - x$$

which is linear and has the solution

$$z = \frac{Cxe^{-x} - 1}{2x}.$$

Thus

$$y = x + \left(\frac{Cxe^{-x} - 1}{2x}\right)^{-1} = \frac{Cx^2 e^{-x} + x}{Cxe^{-x} - 1} = \frac{x\,(Cx + e^x)}{Cx - e^x}. \qquad \square$$

1.9   SINGULAR SOLUTIONS

By choosing a value for the arbitrary constant C in the general solution (1.2), a particular solution of the differential equation is obtained. Solutions that are not accessible by this route are called **singular solutions**.

Example 1.8   Solve $y' = 3xy^{1/3}$, $y(0) = 0$.

Solution 1.8   Separation of the variables leads to the general solutions

$$y = \pm \left(x^2 + C\right)^{3/2}.$$

From the initial condition, $C = 0$ and thus both $y = x^3$ and $y = -x^3$ are particular solutions. However, $y \equiv 0$ is also a solution! This solution is not accessible to the separation of variables procedure which explicitly assumes that it is permissible to divide by $y^{1/3}$, that is, y is at worst zero at discrete points and not over an interval of finite length. Thus this initial value problem has a non–unique solution. $\qquad \square$

1.10   TUTORIAL EXAMPLES 1

Find the general solution of the following first order ordinary differential equations.

T 1.1 $y' - y = \sin 2x.$

T 1.2 $x^2 y' + xy + 1 = 0.$

T 1.3 $y' + (1 - y^2) \tan x = 0.$

T 1.4 $x(x^2 - 6y^2) y' = 4y(x^2 + 3y^2).$

T 1.5 $(3x^2 + 4xy) + (2x^2 + 3y^2) y' = 0.$

T 1.6 $y^2 + (xy + 1) y' = 0.$

T 1.7 $xy' = y + 2xy^2.$

T 1.8 $(x^2 + a) y' + 2y^2 - 3xy - a = 0.$

T 1.9 $(x^2 + 1) y' - xy = x^3 + x.$

Solve the following initial value problems.

T 1.10 $y' + 2xy = e^{-x^2}, \quad y(0) = 1.$

T 1.11 $y' + (2/x)y = (\cos x)/x^2, \quad y(\pi) = 0 \ (x > 0).$

T 1.12 $y' - xy/2 = xy^5, \quad y(0) = a.$

T 1.13 $x^2 y - (x^3 + ay^3) y' = 0, \quad y(0) = 1.$

T 1.14 $y' - 2y = x^2 e^{2x}, \quad y(0) = 0.$

T 1.15   Solve the initial value problem

$$(x - 2y) y' = 2x - y, \qquad y(1) = 3,$$

find an explicit expression for $y = y(x)$ and determine the range of x for which the solution is valid.

2

Modelling Applications

In this chapter we shall encounter a variety of different problems arising in applied mathematics, physics, engineering and bioscience. Our objective with these problems will be to formulate a simple mathematical model and solve it accordingly.

2.1 NEWTON'S LAW OF COOLING

Newton's Law of cooling states that the rate of heat loss from the surface of an object is proportional to the difference in temperature between the object's surface and its environment. If the object is a very good conductor of heat (often true in practice) then the internal temperature of the body is effectively that of its surface. Under these conditions, Newton's Law of cooling claims that the rate of change in the temperature of an object is proportional to the difference in temperature between the object and its environment. Let $T(t)$ and $S(t)$ be the temperatures of the object and its environment at time t respectively. Newton's Law of cooling states that

$$\frac{dT}{dt} = -k(T - S) \tag{2.1}$$

where the proportionality in the statement of the law has been changed to an equality by the introduction of a constant k (physical dimension: time^{-1}). Equation (2.1) is rewritten as[1]

$$\dot{T}(t) + kT(t) = kS(t) \tag{2.2}$$

---

[1] In this section and on many occasions thereafter we shall denote the independent variable by t, simply because in a significant number of applications the independent variable is *time*. Derivatives with respect to t are indicated by dots: $\dot{f}(t) = df(t)/dt$, $\ddot{f}(t) = d^2 f(t)/dt^2$.

The Gompertz law

Living organisms provide one of the best examples of populations. Often neither the Malthusian nor logistic models of population growth are appropriate for biological organisms. One such model assumes that cell populations $P(t)$ evolve according to the **Gompertz equation**

$$\frac{dP}{dt} = (a - b \ln P)P. \tag{2.11}$$

With a and b both positive numbers, it is clear that this model predicts a maximum population size of $e^{a/b}$. The law (2.11) was first proposed by Gompertz in the 19th century as a model for the force of human mortality, $\mu(t)$, at age t. It was intended to reflect more accurately the increasing risk of death as individuals aged but still remain sufficiently tractable to enable numerous actuarial calculations to be done analytically. To be specific, Gompertz proposed the formula

$$\mu(t) = Ae^{c^t} \tag{2.12}$$

where A and c are positive constants. Clearly $\ln \mu = \ln A + c^t$ and therefore

$$\frac{1}{\mu}\frac{d\mu}{dt} = (\ln c)\, c^t = (\ln \mu - \ln A) \ln c$$

which may in turn be re–expressed as

$$\frac{d\mu}{dt} = \left[(\ln c)(\ln \mu) - (\ln A)(\ln c) \right] \mu. \tag{2.13}$$

Thus μ, the force of mortality at age t, is seen to fulfil a differential equation of type (2.11) (provided $0 < c < 1$ and $A > 1$), although Gompertz did not directly propose that μ should satisfy a differential equation.

Solutions of (2.11) are obtained by changing the dependent variable from P to $\phi = \ln P$. Thereby it is seen that ϕ satisfies the linear equation

$$\frac{d\phi}{dt} = a - b\,\phi. \tag{2.14}$$

Solutions to the Gompertz equation are explored in more detail in Tutorial example T 2.4.

2.3   SIMPLE MOTION FROM NEWTON'S SECOND LAW

Newton's Second Law states that the rate of change of the momentum of an object (or a particle) equals all external forces acting on that object, that is,

$$\frac{dp}{dt} = F_{\text{ext}} \tag{2.15}$$

where p is the **momentum** and F_{ext} contains all the external forces. In non–relativistic mechanics the momentum of a particle can be written as

$$p = mv \qquad (2.16)$$

where m is its mass and v is its velocity. It is important to realize that momentum is a vector quantity. For the purposes of this book, however, we shall often confine ourselves to one–dimensional motion. In this case, momentum and velocity are algebraic (they have magnitude and sign).

Remarks 2.2   One–dimensional motion can be described by

$$\text{position :} \qquad x = x(t)$$
$$\text{velocity :} \qquad v(t) = dx/dt = \dot{x}$$
$$\text{acceleration :} \qquad a(t) = dv/dt = \dot{v} = d^2x/dt^2 = \ddot{x}\,.$$

We also remark that the term *speed* refers to the absolute value of the velocity: $|v| = \sqrt{\dot{x}^2 + \dot{y}^2}$. $\triangle$

With these preparations, the one–dimensional motion of a solid body influenced by **gravity** and **air resistance** is now studied. As discussed, define $v(t)$ to be the velocity of the body at time t. Gravity is assumed to be a constant external force. Let the coordinate system be chosen such that the force of gravity acts in the negative x direction. In this instance $F_{\text{g}} = -mg$ where $g = 9.81$ m/s^2 is the gravitational acceleration. Air resistance can often be modelled as an external force whose magnitude is proportional to a power of the speed but which acts in the opposite direction to v, that is,

$$F_{\text{res}} = -k|v|^{n-1}\, v\,. \qquad (2.17)$$

The actual physical dimension of the constant $k > 0$ depends on the value of n. For a linear power law ($n = 1$) for the air resistance, Newton's Second Law states

$$\frac{d}{dt}(mv) = F_{\text{g}} + F_{\text{res}} = -mg - mkv\,. \qquad (2.18)$$

Note that the proportionality in the air resistance term has been re–expressed in (2.18) by $-mkv$ instead of, as suggested by (2.17), $-kv$ so that mass m can be cancelled conveniently from the whole equation. Consequently, the physical dimension of k is now time^{-1}. The *initial value problem* is then

$$\frac{dv}{dt} = -g - kv\,, \qquad v(t_0) = v_0 \qquad (2.19)$$

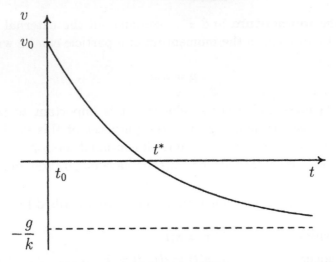

Figure 2.3: Velocity v as a function of time t as per (2.20). At time t^* the body reaches the maximum height of its trajectory.

where v_0 is the initial velocity. The particular solution is

$$v(t) = \left(\frac{g}{k} + v_0 \right) e^{-k(t-t_0)} - \frac{g}{k}. \tag{2.20}$$

While (2.20) is valid for all v_0, we shall now concentrate on $v_0 > 0$ for which the velocity profile (2.20) is displayed in Figure 2.3. The time t^* is defined as $v(t^*) = 0$ and corresponds to a change in the sign of v; in other words, the rising motion of the object has come to an end and will be followed by its fall. It is straightforward to show from (2.20) that

$$t^* = t_0 + \frac{1}{k} \ln \left(1 + \frac{k v_0}{g} \right). \tag{2.21}$$

The position of the object with initial position $x(t_0) = x_0$ follows from

$$x(t) = x_0 + \int_{t_0}^{t} v(s)\, ds = x_0 + \left(\frac{g}{k} + v_0 \right) \frac{1 - e^{-k(t-t_0)}}{k} - \frac{g}{k}(t - t_0). \tag{2.22}$$

The maximum height $x_{\max}$ which is reached at $t = t^*$ is

$$x_{\max} = x(t^*) = x_0 + \frac{v_0 - g\,(t^* - t_0)}{k}. \tag{2.23}$$

It is apparent from (2.20) and Figure 2.3 that $v(t) \to -g/k$ as $t \to \infty$. The velocity g/k is called the **terminal velocity**. Note also that the rate of change of v, i.e., the acceleration itself, approaches zero as $t \to \infty$.

2.4   CURVES OF PURSUIT

At time $t = 0$, a fox at $F(a, 0)$ spots at the origin $R(0, 0)$ a rabbit running at a constant speed v in the positive y direction. The fox runs at constant speed w in a path that is always directed at the rabbit. What is the fox's path and where (if at all) does the fox catch the rabbit?

Let $y = y(x)$ be the path of the fox. Since the fox always runs towards the rabbit, then, as shown in Figure 2.4, the tangent to the fox's path at any time intersects the y axis at the position of the rabbit. Thus

$$y' = \frac{-(vt - y)}{x} = \frac{y - vt}{x}$$

or

$$xy'(x) = y(x) - vt. \tag{2.24}$$

This differential equation is not yet in a form in which it can be solved for $y(x)$ because it contains time t explicitly (essentially because the problem describes two–dimensional *motion*). More information is required. The distance s run by the fox may be expressed in terms of the **arc length** of its path by

$$wt = \int ds = \int_x^a \sqrt{1 + y'^2}\, dx$$

where ds is the infinitesimal arc length. Thus

$$t = \frac{1}{w} \int_x^a \sqrt{1 + y'^2}\, dx. \tag{2.25}$$

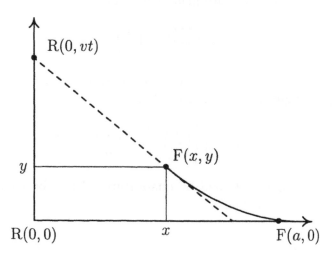

Figure 2.4: Geometry of the pursuit problem. The initial positions of the fox and the rabbit are at F(a,0) and R(0,0) respectively.

This expression is now used to eliminate t from (2.24) to obtain

$$xy' - y = -\frac{v}{w} \int_x^a \sqrt{1 + y'^2}\, dx\,. \tag{2.26}$$

The last expression looks even less like a standard differential equation. Indeed, it is called an **integro–differential** equation because it contains a mixture of integrals and derivatives involving the unknown function. However, on differentiating both sides with respect to x and using the fundamental theorem of calculus from Remarks 1.1 it follows that

$$xy'' = \frac{v}{w}\sqrt{1 + y'^2}\,. \tag{2.27}$$

This is a conventional differential equation of second order.[2]

Remarks 2.3   A differential equation of second order in which the dependent variable y does not explicitly appear, i.e., $F[x, y', y''] = 0$, can be reduced to a first order equation by using $p = y'$ as a new dependent variable. The new form $F[x, p, p'] = 0$ is recognized as the general first order equation (1.1) which can be solved for p. The original unknown y is then determined by *quadrature* as $y = \int p\, dx$. $\triangle$

Now let $p = y'$ and $p' = y''$ in (2.27), then

$$xp' = \frac{v}{w}\sqrt{1 + p^2}\,. \tag{2.28}$$

This equation is *separable* with general solution

$$p + \sqrt{1 + p^2} = \left(\frac{x}{a}\right)^{v/w}\,. \tag{2.29}$$

The particular solution for which $y' = p = 0$ at $x = a$ (see the geometry of the starting configuration in Figure 2.4) is

$$\frac{dy}{dx} = p = \frac{1}{2}\left[\left(\frac{x}{a}\right)^{v/w} - \left(\frac{x}{a}\right)^{-v/w}\right]\,. \tag{2.30}$$

The path $y(x)$ is now obtained by quadrature. If the fox runs faster than the rabbit (i.e. $w > v$) then

$$y(x) = \frac{a}{2}\left[\frac{(x/a)^{1+v/w}}{1 + v/w} - \frac{(x/a)^{1-v/w}}{1 - v/w}\right] + B\,. \tag{2.31}$$

---

[2]Differential equations of second order will be treated in detail in Chapters 3–6.

The integration constant B follows from the condition $y(a) = 0$ (the fox's starting point) and leads to $B = avw/(w^2 - v^2)$. Thus, the fox catches the rabbit where its path intersects the y axis, i.e., where $x = 0$. From (2.31) this occurs where

$$y(0) = \frac{avw}{w^2 - v^2}.\qquad(2.32)$$

Thus the fox catches the rabbit at position $y(0)$ after $aw/(w^2 - v^2)$ units of time.

Remarks 2.4   Another way to solve (2.28) is by using the substitution $p = \sinh z$. This yields

$$p = \sinh\left(\frac{v}{w}\ln x + C\right)\qquad\longrightarrow\qquad p = \frac{1}{2}\left[(Ax)^{v/w} - (Ax)^{-v/w}\right]$$

where A and C are integration constants and we have used the definition of sinh in terms of exponential functions. The 'initial' condition $p = 0$ at $x = a$ leads again to (2.30). $\triangle$

2.5   SIMPLE ELECTRICAL CIRCUITS

Simple **electrical circuits** consist of an arrangement of various electrical elements such as, for example, a source of electromotive force — often a battery — and resistors, inductors and capacitors. The basic quantity is the **electrical charge** Q. The **electrical current** flowing in a circuit is denoted by I and is derived from Q via

$$I(t) = \frac{dQ(t)}{dt}.\qquad(2.33)$$

Current is therefore the rate of flow of charge. Circuit elements experience potential drops/differences denoted by U (**electrical voltage**). Table 2.1 lists simple circuit elements we shall be dealing with and describes their properties.[3]

The behaviour of electrical circuits is based on **Kirchhoff's Laws**. There are two laws, the first of which simply states that current is conserved at any junction (point at which wires connect). The method of *loop currents* introduces dependent variables in such a way that Kirchhoff's First Law is satisfied automatically. It is therefore the second of Kirchhoff's Laws that provides the differential equations. Here Kirchhoff's Second Law is expressed

---

[3]The reader should be aware that in problems involving electrical circuits that contain capacitors, the symbol C is used for the capacitance and not, as elsewhere in the book, as the standard symbol for an integration constant.

rabbit. Find and discuss the fox's path for the case where the fox does not run faster than the rabbit ($w \leq v$). For the case of equal velocity ($w = v$) find the closest distance between the fox and the rabbit.

T 2.9   More curves of pursuit. A research vessel is able to locate a whale on the ocean surface some 4 km away. The whale dives instantly and proceeds at full speed in an unknown direction. What path should the ship select to be certain of passing directly over the whale (at some time) if its velocity is 3 times that of the whale.
[*Hint*: use polar coordinates r, θ and $ds^2 = dr^2 + r^2 d\theta^2$ to delineate the arc length. Also, suppose that the ship proceeds in the following way: initially, for a period of time in a straight line aimed directly at the point where the whale was spotted, and then in a curve described by $r = r(\theta)$.]

T 2.10   Electrical circuits. A closed circuit consists of an electromotive force (a battery) of voltage output $U(t)$ connected in series to a capacitor of capacitance C and a resistor of resistance R. Model this circuit and solve the differential equation for the charge on the capacitor for each of the following problems.

1. Assume that the capacitor is charged to $Q = Q_0$ at time $t = t_0$ when the circuit is activated. Assuming that the circuit has no battery, explain what happens by interpreting the solution of the differential equation under these initial conditions.

2. Assume that the capacitor is initially discharged, i.e., $Q = 0$ at time $t = t_0$ when the circuit is activated. A battery of constant voltage output $U = U_0$ is present. How long does it take until the capacitor is charged to half its maximum value?

3. For the sinusoidal electromotive force $U(t) = U_0 \sin \omega t$, investigate the behaviour of the voltage at the capacitor. Discuss the solution. What happens as $t \to \infty$?

3

Linear Differential Equations of Second Order

3.1 MECHANICAL AND ELECTRICAL SYSTEMS

Linear second order differential equations, particularly those with constant coefficients, occur widely in disciplines which use mathematics for modelling purposes. Probably the most common applications arise in the discussion of the concept of resonance which is usually produced by some external periodic input, for example, wind–inducing oscillations in a bridge or excessive rolling of ships due to wind and wave action. Of course, not all resonance effects are undesirable. For example, most radio and television receivers use tuned circuits. These are essentially resonators whose natural frequency can be adjusted.

Consider a one–dimensional system whose deviation from its equilibrium position at time t is described by $y(t)$ and whose motion is influenced by:

- a restoring force directly proportional to y but in opposition to it (because it is a restoring influence);

- a damping force directly proportional to $\dot{y}$ but opposing it;

- an external driving force modelled by a known function $f(t)$.

In a typical mechanical system, $y(t)$ is a displacement, damping is produced by dashpots/shock absorbers and restoring forces are due to spring–like components within the system. In a fluid system, resistance is typically due to viscosity with restoring forces generated by buoyancy effects (due to gravity).

The physical description of a mechanical model or a fluid model comes from Newton's Second Law (see Section 2.3) which asserts that the rate of change of momentum of the system is equal to the sum of the applied forces. This leads naturally to the differential equation[1]

$$m\ddot{y} = -b\dot{y} - ky + f(t) \qquad (3.1)$$

where m is mass (assumed constant), b is the damping parameter, k is the spring/buoyancy parameter and $f(t)$ is an external force. Both b and k are positive in realistic problems but, of course, the differential equation exists as a mathematical entity for all real (as well as complex) values of the constants.

The primitive concept in electromagnetism[2] is electrical charge $Q(t)$ (measured in *Coulomb*) and is formally equivalent to displacement in mechanics — all electrical entities are spawned from charge through differentiation just as differentiation spawns mechanical entities from displacement. Electrical current $I(t)$ (measured in *Ampere*) measures the flow or rate of change of charge and is formally equivalent to velocity in mechanics, that is, $I(t) = \dot{Q}(t)$. Electrical components which damp or dissipate charge are resistors (commonly denoted by the symbol R). Just as springs store mechanical energy by virtue of displacement, capacitors (commonly denoted by the symbol C) store electrical energy by virtue of charge. The electrical equivalent of mechanical force is called voltage difference (voltage is commonly denoted by the symbol U) — voltage difference is analogous to electrical force. The effect of inertia is provided by an inductor (commonly denoted by the symbol L) and is the electrical equivalent of mass, see also Table 2.1. Just as Newton's Second Law states that externally applied forces are balanced by $(d/dt)\, mv$ then *Kirchhoff's Second Law* (see also Section 2.5) states that the sum of all voltage drops around a closed circuit is zero. Applied to the LCR circuit shown in Figure 3.1 we therefore get

$$L\dot{I}(t) + RI(t) + \frac{Q(t)}{C} = U(t) \qquad (3.2)$$

where L is the inductance, R is the resistance, C is the capacitance and $U(t)$ is the electromotive force. Recalling $I = \dot{Q}$, (3.2) is recognized as a differential equation of second order for the charge Q on the capacitor.

3.2   GENERAL CONSIDERATIONS

Both of the introductory examples of the previous section give rise to differential equations of the type

$$\ddot{y} + 2\alpha\dot{y} + \omega^2 y = f(t) \qquad (3.3)$$

---

[1]We mentioned previously that such problems have in general vector character.
[2]See also Section 2.5.

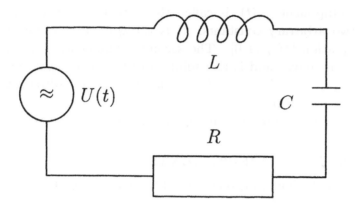

Figure 3.1: Electrical circuit connecting — in series — an electromotive force $U(t)$, resistor R, capacitor C and inductor L.

for a function $y = y(t)$ and where α is a positive parameter related to system damping. If no external driving mechanism is present then $f(t) \equiv 0$ and the corresponding differential equation is said to be **homogeneous**, otherwise it is **inhomogeneous**. Special attention will be devoted to second order linear differential equations with positive coefficients, but it is appropriate to begin with the more general equation

$$\frac{d^2y}{dt^2} + a\frac{dy}{dt} + by = f(t) \qquad (3.4)$$

where a and b are complex–valued constants. The differential equation (3.4) is a special case of the most general, linear, inhomogeneous, second order ordinary differential equation

$$\frac{d^2y}{dt^2} + a(t)\frac{dy}{dt} + b(t)y = f(t). \qquad (3.5)$$

Usually solutions of (3.5) are sought for $t \geq t_0$ such that the conditions

$$y(t_0) = y_0, \qquad \dot{y}(t_0) = \dot{y}_0 \qquad (3.6)$$

are satisfied where y_0 and $\dot{y}_0$ are given constants. The equation (3.5) and conditions (3.6) specify an **initial value problem of second order**.

The **general solution** of (3.5) can be written as the sum of two components, that is, in the form

$$y(t) = y_c(t) + y_p(t). \qquad (3.7)$$

may then be rewritten as

$$y(t) = e^{\eta t} \left(C \cosh \xi t + D \sinh \xi t \right)$$

by virtue of the definitions of the sinh and cosh functions.   △

Of course, there is an element of dishonesty in the solution process of this section in the sense that solutions to these equations are obtained without any integration whatsoever. Some loose ends to be tidied up are:

- Is there an explanation for the special form of the solution (3.13) which arises when the auxiliary equation has a repeated root?

- Is there a general method for solving inhomogeneous equations when a forcing term is present?

- Forcing terms of type $f(t) = P(t)e^{\lambda t} \sin t$ or $f(t) = P(t)e^{\lambda t} \cos t$, where $P(t)$ is a polynomial in t, can be treated systematically without integration provided certain, somewhat obscure rules are obeyed, see Section 4.2. No other functions $f(t)$ are amenable to this approach.

3.4   SOLUTION OF THE INHOMOGENEOUS EQUATION

This section is directed at a more coherent (but also considerably more involved) treatment of inhomogeneous second order differential equations with constant coefficients. The ideas can be extended to higher order equations. Consider the second order equation

$$\ddot{y}(t) + a\dot{y}(t) + by(t) = f(t), \qquad t > t_0, \tag{3.14}$$

where a, b are complex–valued constants and f is an arbitrary function of t.

Let $\lambda_1 = \alpha$ and $\lambda_2 = \beta$ be the possibly complex–valued roots of the auxiliary equation $\lambda^2 + a\lambda + b = 0$ then, in terms of α and β, the differential equation assumes the form

$$\left(\frac{d}{dt} - \alpha \right) \left(\frac{dy}{dt} - \beta y \right) = f(t) \tag{3.15}$$

when written in operator notation.[4] Now define $z = \dot{y} - \beta y$ so that the original second order equation reduces to the two first order linear equations

$$\frac{dz}{dt} - \alpha z = f(t), \qquad \frac{dy}{dt} - \beta y = z \tag{3.16}$$

---

[4]When evaluating an operator equation of the type (3.15), one must remember that the differential operator d/dt acts on all functions of t to its right.

for $y(t)$ and $z(t)$. The two equations (3.16) form what is generally called a **system of first order differential equations** (which will be treated in more detail in Chapter 8). In this particular case, (3.16) can be solved in sequence for z first and then for y. Clearly

$$\frac{dz}{dt}e^{-\alpha t} - \alpha z e^{-\alpha t} = f(t)e^{-\alpha t} \qquad \longrightarrow \qquad \frac{d(ze^{-\alpha t})}{dt} = f(t)\,e^{-\alpha t}$$

and this integrates to

$$z(t) = e^{\alpha t}\int_{t_0}^{t} e^{-\alpha s}f(s)\,ds + Ae^{\alpha t}\,. \qquad (3.17)$$

where A is an arbitrary complex–valued constant. Once $z(t)$ is known then $y(t)$ can be determined by a further integration. The integrating factor in this case is $e^{-\beta t}$ and the equation for $y(t)$ simplifies to

$$\frac{d(ye^{-\beta t})}{dt} = e^{(\alpha-\beta)t}\int_{t_0}^{t} e^{-\alpha s}f(s)\,ds + Ae^{(\alpha-\beta)t}\,.$$

Hence the final solution for $y(t)$ is

$$y(t) = e^{\beta t}\int_{t_0}^{t} e^{(\alpha-\beta)u}\int_{t_0}^{u} e^{-\alpha s}f(s)\,ds\,du + Ae^{\beta t}\int_{t_0}^{t} e^{(\alpha-\beta)u}\,du + Ce^{\beta t} \quad (3.18)$$

where C is a second arbitrary complex–valued constant. It is now clear why the case $\alpha = \beta$ causes difficulty — it is because of the changing nature of $e^{(\alpha-\beta)u}$ in the integrand of (3.18). Each case is treated in turn.

Case 1: $\alpha = \beta$. In this situation, the previous expression for y simplifies to

$$
\begin{aligned}
y(t) &= e^{\beta t}\int_{t_0}^{t}\int_{t_0}^{u} e^{-\beta s}f(s)\,ds\,du + Ae^{\beta t}\int_{t_0}^{t} du + Ce^{\beta t}\\
&= e^{\beta t}\left[u\int_{t_0}^{u} e^{-\beta s}f(s)\,ds\right]_{t_0}^{t} - e^{\beta t}\int_{t_0}^{t} ue^{-\beta u}f(u)\,du\\
&\qquad + Ae^{\beta t}(t-t_0) + Ce^{\beta t}\\
&= -e^{\beta t}\int_{t_0}^{t}(u-t)e^{-\beta u}f(u)\,du + Ate^{\beta t} + (C-t_0 A)e^{\beta t}\\
&= \int_{t_0}^{t}(t-u)e^{\beta(t-u)}f(u)\,du + (At+B)e^{\beta t}
\end{aligned}
$$

where integration by parts has been used. When the substitution $s = t - u$ is used to change the integration variable from u to s, the result is

$$y(t) = (At+B)e^{\beta t} + \int_{0}^{t-t_0} f(t-s)se^{\beta s}\,ds\,. \qquad (3.19)$$

Case 2: $\alpha \neq \beta$. In this case, the expression (3.18) for y becomes

$$y(t) = e^{\beta t} \int_{t_0}^{t} e^{(\alpha-\beta)u} \int_{t_0}^{u} e^{-\alpha s} f(s)\, ds\, du + Ae^{\beta t} \left[\frac{e^{(\alpha-\beta)u}}{\alpha-\beta} \right]_{t_0}^{t} + Ce^{\beta t}$$

$$= e^{\beta t} \int_{t_0}^{t} e^{(\alpha-\beta)u} \int_{t_0}^{u} e^{-\alpha s} f(s)\, ds\, du + De^{\alpha t} + Ee^{\beta t}$$

where D and E are arbitrary constants. As in the previous case, the double integral is evaluated by parts to obtain

$$e^{\beta t} \left[\frac{e^{(\alpha-\beta)u}}{\alpha-\beta} \int_{t_0}^{u} e^{-\alpha s} f(s)\, ds \right]_{t_0}^{t} - e^{\beta t} \int_{t_0}^{t} \frac{e^{(\alpha-\beta)u}}{\alpha-\beta} e^{-\alpha u} f(u)\, du \, .$$

A similar effect can be achieved by change of order in the double integral. As a result, the general solution can now be expressed in the form

$$y(t) = -e^{\beta t} \int_{t_0}^{t} \frac{e^{(\alpha-\beta)u} - e^{(\alpha-\beta)t}}{\alpha-\beta} e^{-\alpha u} f(u)\, du + Ae^{\alpha t} + Be^{\beta t}$$

$$= -\frac{e^{\beta t}}{\alpha-\beta} \int_{t_0}^{t} (e^{-\beta u} - e^{\alpha(t-u)} e^{-\beta t}) f(u)\, du + Ae^{\alpha t} + Be^{\beta t}$$

$$= \frac{1}{\alpha-\beta} \int_{t_0}^{t} (e^{\alpha(t-u)} - e^{\beta(t-u)}) f(u)\, du + Ae^{\alpha t} + Be^{\beta t}$$

$$= \frac{1}{\alpha-\beta} \int_{0}^{t-t_0} (e^{\alpha s} - e^{\beta s}) f(t-s)\, du + Ae^{\alpha t} + Be^{\beta t} \, . \qquad (3.20)$$

Let $\Phi(\lambda, t)$ be defined by the integral

$$\Phi(\lambda, t) = \int_{0}^{t-t_0} e^{\lambda s} f(t-s)\, ds \qquad (3.21)$$

then we see from (3.19), (3.20) that $\ddot{y} + a\dot{y} + by = f(t)$ has general solution

$$y(t) = \begin{cases} (At+B)e^{\alpha t} + \dfrac{d\Phi(\lambda, t)}{d\lambda}\bigg|_{\lambda=\alpha} & \alpha = \beta, \\[4mm] Ae^{\alpha t} + Be^{\beta t} + \dfrac{\Phi(\alpha, t) - \Phi(\beta, t)}{\alpha-\beta} & \alpha \neq \beta. \end{cases} \qquad (3.22)$$

In the expressions (3.22), the terms comprising the complementary function, see equations (3.11) and (3.13), and the particular integral are readily recognizable.

Example 3.4 Find a particular integral of $\ddot{y} + 4\dot{y} + 3y = 2e^t$ for $t \geq 0$.

Solution 3.4   The auxiliary equation is $\lambda^2 + 4\lambda + 3 = (\lambda + 1)(\lambda + 3) = 0$ with roots $\alpha = -1$, $\beta = -3$ and thus $y_c(t) = Ae^{-t} + Be^{-3t}$. Hence

$$\Phi(\lambda, t) = \int_0^t 2e^{\lambda s} e^{t-s}\, ds = 2e^t \int_0^t e^{(\lambda - 1)s}\, ds = 2e^t \frac{e^{(\lambda - 1)t} - 1}{\lambda - 1}$$

$$= \frac{2\left(e^{\lambda t} - e^t\right)}{\lambda - 1}.$$

The particular integral is

$$\frac{\Phi(-1, t) - \Phi(-3, t)}{(-1) - (-3)} = \frac{1}{2}\left[\left(e^t - e^{-t}\right) - \frac{1}{2}\left(e^t - e^{-3t}\right)\right].$$

It is clear that the second and the last term in this expression are already constituents of the complementary function y_c and so they can be ignored safely. A suitable particular integral is therefore

$$y_p(t) = \frac{1}{2}\left(e^t - \frac{e^t}{2}\right) = \frac{e^t}{4}.$$

Compare this with the standard approach in the method of undetermined coefficients (see Section 4.2) where Ae^t is pursued as a particular integral for a suitable choice of A. In this case, the current method is clearly more involved but both produce the same particular integral.   □

Example 3.5   Find a particular integral for the inhomogeneous equation $\ddot{y} + y = t\sin t$ valid in the interval $t \geq 0$.

Solution 3.5   Here the auxiliary equation is $\lambda^2 + 1 = 0$ with roots $\alpha = i$, $\beta = -i$ and thus $y_c(t) = A\cos t + B\sin t$. In this case, a particular integral is

$$y_p(t) = \frac{\Phi(i, t) - \Phi(-i, t)}{\alpha - \beta} = \frac{1}{2i}\int_0^t (e^{is} - e^{-is})(t - s)\sin(t - s)\, ds$$

$$= \frac{1}{2i}\int_0^t (2i\sin s)(t - s)\sin(t - s)\, ds$$

$$= \int_0^t (t - s)\sin s\sin(t - s)\, ds$$

$$= \frac{1}{2}\int_0^t \left[(t - s)\sin s\sin(t - s) + s\sin(t - s)\sin s\right] ds$$

$$= \frac{t}{2}\int_0^t \sin s\sin(t - s)\, ds = -\frac{t}{4}\int_0^t \left[\cos t - \cos(t - 2s)\right] ds$$

$$= -\frac{1}{4}(t^2 \cos t - t\sin t).$$ □

4.2   THE METHOD OF UNDETERMINED COEFFICIENTS

The method of **undetermined coefficients** is based on a guess for the form of y_p. The technique can be used if the function $g(x)$ on the right–hand side of (4.1) is limited to terms containing sums of products of polynomials with exponential functions, sines and cosines (all with linear arguments). To understand the strategy, let $\hat{H}$ be the **linear differential operator** defined by

$$\hat{H}(y) = a\frac{d^2y}{dx^2} + b\frac{dy}{dx} + cy \tag{4.3}$$

and let $p(\lambda) = a\lambda^2 + b\lambda + c$, then the auxiliary equation associated with $\hat{H}$ is $p(\lambda) = 0$. For any complex–valued constant σ and suitably differentiable function $h(x)$, it can be shown that

$$\hat{H}\left[h(x)e^{\sigma x}\right] = \left[a\frac{d^2h}{dx^2} + (2a\sigma + b)\frac{dh}{dx} + (a\sigma^2 + b\sigma + c)h\right]e^{\sigma x}. \tag{4.4}$$

For any polynomial $q(x)$, $h(x)e^{\sigma x}$ is a particular integral of $\hat{H}(y) = q(x)e^{\sigma x}$ provided

$$a\frac{d^2h}{dx^2} + (2a\sigma + b)\frac{dh}{dx} + (a\sigma^2 + b\sigma + c)h = q(x). \tag{4.5}$$

There are clearly three different cases to consider.

Case 1: If $p(\sigma) \neq 0$, that is, σ is not a root of the auxiliary equation of the operator $\hat{H}$, (4.5) is satisfied by an appropriate polynomial $h(x)$ whose degree is the same as that of $q(x)$.

Case 2: If σ is now a *single* root of the auxiliary equation $p(\lambda) = 0$, then it may be verified that $a\sigma^2 + b\sigma + c = 0$ but $2a\sigma + b \neq 0$. Therefore (4.5) now assumes the simplified form

$$a\frac{d^2h}{dx^2} + (2a\sigma + b)\frac{dh}{dx} = q(x), \tag{4.6}$$

which in turn integrates to the expression

$$a\frac{dh}{dx} + (2a\sigma + b)h = \int^x q(s)\,ds. \tag{4.7}$$

This relationship may be satisfied by an appropriate polynomial h of degree one more than that of q. In particular, the constant term in h plays no role in determining q and may be set to zero without loss of generality. Thus $h(x) = xr(x)$ where r is a polynomial in x with the same degree as q.

Case 3: Finally, if σ is a double root of the auxiliary equation $p(\lambda) = 0$ then $p(\sigma) = 0$ and $p'(\sigma) = 2a\sigma + b = 0$ in (4.5). Thus

$$a\frac{d^2h}{dx^2} = q(x) \tag{4.8}$$

and so particular integrals are now polynomials $h(x)$ of degree two more than that of $q(x)$. Furthermore, it is clear from (4.5) that the constant and coefficient of x in $h(x)$ play no role in the determination of h from q and so there is no loss in generality in setting both these coefficients to be zero. Thus $h(x) = x^2 r(x)$ where r is a polynomial in x whose degree is that of q.

This completes the discussion of the choice of particular integrals for second order linear differential equations with constant coefficients. However, this theme will be revived in more generality during the discussion of linear differential equations of higher order with constant coefficients (see Section 7.3).

These ideas are now applied to find particular integrals for inhomogeneous second order linear differential equations. If

$$P_n(x) = a_0 + a_1 x + \ldots + a_n x^n$$

is a (given) polynomial of degree n, then Table 4.1 lists the form of y_{p_i} necessary to find a particular integral for a given g_i.

$g_i(x)$	$y_{pi}(x)$
$P_n(x)$	$x^s Q_n(x)$
$P_n(x)\,e^{\alpha x}$	$x^s Q_n(x)\,e^{\alpha x}$
$P_n(x)\,e^{\alpha x} \sin \beta x$	$x^s e^{\alpha x} \left[Q_n(x) \sin \beta x + R_n(x) \cos \beta x \right]$
$P_n(x)\,e^{\alpha x} \cos \beta x$	$x^s e^{\alpha x} \left[Q_n(x) \sin \beta x + R_n(x) \cos \beta x \right]$

Table 4.1: Solution strategy for the method of undetermined coefficients.

In Table 4.1,

$$Q_n(x) = A_0 + A_1 x + \ldots + A_n x^n\,,$$
$$R_n(x) = B_0 + B_1 x + \ldots + B_n x^n\,,$$

are two polynomials of degree n with as yet undetermined coefficients A_j, B_j, $j = 1, 2, \ldots, n$. The exponent s is chosen from the set $\{0,1,2\}$ where s is the **multiplicity** of $\alpha + i\beta$ in the auxiliary equation $p(\lambda) = 0$: $s = 0$ when $\alpha + i\beta$ is not a root, $s = 1$ when $\alpha + i\beta$ is a single root and $s = 2$ when $\alpha + i\beta$ is a double root of $p(\lambda) = 0$. Upon substitution of $y_{p_i}(x)$ into the differential equation, the unknown coefficients can be completely determined by **comparison of coefficients**.

Remarks 4.1   The four different cases which were distinguished in Table 4.1 are all contained in $g_i(x) = P_n(x)\,e^{\alpha x}$, $y_{pi}(x) = x^s Q_n(x)\,e^{\alpha x}$ whereby α is permitted to be complex–valued.   △

Example 4.1   Find a particular integral of the second order differential equation $y'' - 3y' - 4y = 2\sin x$.

Solution 4.1   The auxiliary equation $\lambda^2 - 3\lambda - 4 = (\lambda + 1)(\lambda - 4) = 0$ has roots $\lambda_1 = -1$ and $\lambda_2 = 4$ so that the complementary function is

$$y_c(x) = c_1 e^{-x} + c_2 e^{4x}.$$

Table 4.1 indicates that the particular integral is

$$y_p(x) = A\sin x + B\cos x$$

for suitable values of A and B. Here $s = 0$ because i and $-i$ are not solutions of the auxiliary equation. Insertion into the differential equation gives

$$(-A + 3B - 4A)\sin x + (-B - 3A - 4B)\cos x = 2\sin x.$$

Comparing the coefficients on the left–hand and right–hand sides yields a linear system of equations for the two unknowns A and B as per

$$
\begin{aligned}
-5A + 3B &= 2 \\
-3A - 5B &= 0
\end{aligned}
\qquad \longrightarrow \qquad
\begin{aligned}
A &= -5/17 \\
B &= 3/17
\end{aligned}
$$

and thus

$$y_p(x) = -\frac{5}{17}\sin x + \frac{3}{17}\cos x.$$

The solution contains both a $\sin x$ and a $\cos x$ term although the inhomogeneous term in the differential equation comprised a $\sin x$ term only.   □

Example 4.2   Find a particular integral of the second order differential equation $y'' - 3y' - 4y = e^{-x}$.

Solution 4.2   As before, the complementary function is

$$y_c(x) = c_1 e^{-x} + c_2 e^{4x}.$$

The right–hand side of the given differential equation meets the specification in Table 4.1 with $P(x)$ a polynomial of degree zero and $s = 1$ since $\alpha = -1$ is a single root of the auxiliary equation. The particular integral is

$$y_p(x) = Axe^{-x}$$

for a suitable value for A. Substitution into the inhomogeneous equation gives

$$\left[(Ax + 3Ax - 4Ax) + (-2A - 3A)\right]e^{-x} = e^{-x}.$$

The first term in the bracket vanishes automatically. Comparing the remaining coefficient produces $A = -1/5$ and therefore

$$y_p(x) = -\frac{1}{5}xe^{-x}.$$ □

Example 4.3   Find an appropriate form for the particular integral of

$$y'' + 4y = x^2 e^{-3x} \sin x - x \sin 2x.$$

Solution 4.3   The complementary function is

$$y_c(x) = c_1 \cos 2x + c_2 \sin 2x.$$

The strategy is now to split the problem into the two sub–problems:

1. $y'' + 4y = x^2 e^{-3x} \sin x$;

2. $y'' + 4y = -x \sin 2x.$

Therefore

1. Since $\alpha = -3 + i$ is not a solution of the auxiliary equation then $s = 0$ and the particular integral is

$$y_{p1}(x) = (A_0 x^2 + A_1 x + A_2)e^{-3x} \cos x + (B_0 x^2 + B_1 x + B_2)e^{-3x} \sin x$$

for suitable values of the constants A_0, A_1, A_2, B_0, B_1 and B_2.

2. In this instance $\alpha = 2i$ is a single solution of the auxiliary equation and so $s = 1$. The correct form for the particular integral is therefore

$$y_{p2}(x) = x(C_0 x + C_1) \cos 2x + x(D_0 x + D_1) \sin 2x.$$

Now, the unknown coefficients can be calculated by comparison of coefficients and finally, the particular integral is $y_p(x) = y_{p1}(x) + y_{p2}(x)$.   □

4.3   PARTICULAR INTEGRALS BY COMPLEX METHODS

Example 4.3 highlights the difficulties that are inherent in using real–valued functions to treat linear differential equations whose inhomogeneous terms contain sine and cosine components, that is, terms that arise from complex values of α. There are three obvious difficulties:

(a) by contrast with analyses involving non–trigonometrical right–hand sides, twice as many constants seem to be needed as might have otherwise been expected;

(b) the presence of sine and cosine terms destroys the symmetry enjoyed by exponents;

(c) the connection between the value of α and roots of the auxiliary equation is obscure.

All of these unsatisfactory points can be remedied by expressing real inhomogeneous terms (right–hand side) as the real part of a complex function. The differential equation may now be treated as a fully complex equation with a complex–valued particular integral. The real particular integral is extracted by taking the real part of the complex particular integral. The analysis of non–trigonometric inhomogeneous terms is unchanged. Trigonometric inhomogeneous terms can be made complex by replacing occurrences of $\cos\beta x$ and $\sin\beta x$ by $e^{i\beta x}$ and $-ie^{i\beta x}$ respectively. The motivation for these choices stems from the observations

$$\cos\beta x = \mathrm{Re}\left(e^{i\beta x}\right), \qquad \sin\beta x = \mathrm{Re}\left(-ie^{i\beta x}\right).$$

The analysis is now conventional except that the particular integral is usually complex although it is only its real part that is relevant.

Example 4.4   Find a particular integral of the differential equation

$$y'' + 4y = x^2 e^{-3x}\sin x - x\sin 2x$$

(which is identical to Example 4.3) by constructing a complex differential equation whose real part is this equation.

Solution 4.4   Since $\sin x = \mathrm{Re}\left(-i\,e^{ix}\right)$ and $\sin 2x = \mathrm{Re}\left(-i\,e^{2ix}\right)$, then an appropriate differential equation is

$$z'' + 4z = -ix^2\,e^{(-3+i)x} + ix\,e^{2ix}$$

where $z(x)$ is a complex function and $y(x) = \mathrm{Re}\left[z(x)\right]$. The problem is now split into the two sub–problems:

1. $z'' + 4z = -ix^2\,e^{(-3+i)x}$;

2. $z'' + 4z = ix\,e^{2ix}$.

We shall consider each in turn.

1. Since $\alpha = -3 + i$ is not a root of the auxiliary equation then the complementary function is $z_{p1} = (ax^2 + bx + c)\, e^{(-3+i)x}$ for suitable values of a, b and c. Thus

$$
\begin{aligned}
z'_{p1} &= (2ax + b)\, e^{(-3+i)x} + (-3 + i)z_{p1}\,, \\
z''_{p1} &= 2a\, e^{(-3+i)x} + 2(2ax + b)(-3 + i)\, e^{(-3+i)x} + (-3+i)^2 z_{p1}\,.
\end{aligned}
$$

The constants a, b and c are required to satisfy identically

$$
2a + 2(2ax + b)(-3 + i) + (12 - 6i)(ax^2 + bx + c) = -ix^2
$$

which yields (after some algebra)

$$
a = \frac{1 - 2i}{30}\,, \qquad b = \frac{9 - 13i}{225}\,, \qquad c = \frac{92 - 119i}{6750}\,.
$$

Then,

$$
\begin{aligned}
y_{p1}(x) &= \mathrm{Re}\,(z_{p1}) = \mathrm{Re}\left[(ax^2 + bx + c)\, e^{(-3+i)x}\right] \\
&= \frac{e^{-3x}}{6750}\left[\left(92 + 270x + 225x^2\right)\cos x \right. \\
&\qquad \left. + \left(119 + 390x + 450x^2\right)\sin x\right].
\end{aligned}
$$

2. In this instance $\alpha = 2i$ which is now a single root of the auxiliary equation. Therefore $s = 1$ and $z_{p2} = x(ax + b)e^{2ix}$. By calculation,

$$
\begin{aligned}
z'_{p2} &= (2ax + b)e^{2ix} + 2iz_{p2}\,, \\
z''_{p2} &= 2ae^{2ix} + 4i(2ax + b)e^{2ix} + (2i)^2 z_{p2}\,.
\end{aligned}
$$

Thus a and b are required to satisfy identically

$$
2a + 4i(2ax + b) = ix\,.
$$

As a consequence, $a = 1/8$ and $b = i/16$. The corresponding particular integral is

$$
y_{p2}(x) = \mathrm{Re}\,(z_{p2}) = \mathrm{Re}\left[\left(\frac{x^2}{8} + \frac{ix}{16}\right)e^{2ix}\right] = \frac{x^2}{8}\cos 2x - \frac{x}{16}\sin 2x\,.
$$

The final particular integral is then $y_p(x) = y_{p1}(x) + y_{p2}(x)$.   □

4.4 VARIATION OF CONSTANTS

The idea behind the technique of **variation of constants** follows closely its application to first order equations (see Section 1.6). Yet it requires an additional step as will be seen below. Let $y_c(x) = c_1 y_1(x) + c_2 y_2(x)$ be a solution of the homogeneous equation, i.e., the complementary function. A particular integral of

$$y''(x) + p(x)y'(x) + q(x)y(x) = g(x) \tag{4.9}$$

can then be obtained by writing

$$y_p(x) = u_1(x)y_1(x) + u_2(x)y_2(x) \,. \tag{4.10}$$

The idea is to substitute (4.10) into (4.9) and determine the unknown functions $u_1(x)$ and $u_2(x)$ accordingly. However, this procedure in itself will not provide enough information: after all, there are two unknown functions $u_1(x)$ and $u_2(x)$ but only one differential equation. Another condition is needed on u_1 and u_2 — one which is arbitrary. Differentiate (4.10) to get

$$y_p' = u_1' y_1 + u_1 y_1' + u_2' y_2 + u_2 y_2' \,. \tag{4.11}$$

Before differentiating a second time, impose the arbitrary condition

$$u_1' y_1 + u_2' y_2 = 0 \tag{4.12}$$

which simplifies (4.11) to

$$y_p' = u_1 y_1' + u_2 y_2' \tag{4.13}$$

and thus

$$y_p'' = u_1' y_1' + u_1 y_1'' + u_2' y_2' + u_2 y_2'' \,. \tag{4.14}$$

The motivation for the arbitrary choice (4.12) is that y_p'' now contains only first derivatives of u_1 and u_2. Substituting the last two relations into (4.9), and noting that both y_1 and y_2 are solutions of the homogeneous version of (4.9) gives

$$u_1' y_1' + u_2' y_2' = g \,. \tag{4.15}$$

The two unknown functions u_1 and u_2 can now be calculated from (4.12) and (4.15). Those two relations are viewed as two linear algebraic equations for u_1' and u_2'. Their solutions are

$$u_1'(x) = -\frac{y_2\, g}{W(y_1, y_2)} \,, \qquad u_2'(x) = \frac{y_1\, g}{W(y_1, y_2)} \tag{4.16}$$

where the short–hand notation

$$W(y_1, y_2) = y_1 y_2' - y_1' y_2 \tag{4.17}$$

has been used. The function W is called the **Wronskian** (or **Wronskian determinant**) — it will appear again in Section 7.1. From (4.16), a simple quadrature leads to

$$u_1(x) = -\int \frac{y_2 \, g}{W(y_1, y_2)} \, dx, \quad u_2(x) = \int \frac{y_1 \, g}{W(y_1, y_2)} \, dx. \tag{4.18}$$

The general solution of (4.9) is therefore

$$y(x) = c_1 y_1(x) + c_2 y_2(x) + u_1(x) y_1(x) + u_2(x) y_2(x). \tag{4.19}$$

Example 4.5   Verify that $y_1(x) = x$ and $y_2(x) = 1/x$ are solutions of the differential equation $x^2 y'' + x y' - y = 0$ and find the general solution of $x^2 y'' + x y' - y = x \ln x$ for $x > 0$.

Solution 4.5   Straightforward differentiation and insertion confirms that y_1 and y_2 are solutions of the homogeneous equation. Rewrite the inhomogeneous equation in the form

$$y'' + \frac{y'}{x} - \frac{y}{x^2} = \frac{\ln x}{x}.$$

Thus a particular integral is

$$y_p(x) = x u_1(x) + \frac{1}{x} u_2(x).$$

Following the procedure explained above, $g(x) = (\ln x)/x$ and from (4.17)

$$W(y_1, y_2) = x \left(-\frac{1}{x^2} \right) - \frac{1}{x} = -\frac{2}{x}.$$

By direct substitution into (4.18), we get

$$u_1(x) = -\int \frac{(1/x)(\ln x)/x}{(-2/x)} \, dx = \frac{1}{2} \int \frac{\ln x}{x} \, dx = \frac{1}{4}(\ln x)^2,$$

$$u_2(x) = \int \frac{x(\ln x)/x}{(-2/x)} \, dx = -\frac{1}{2} \int x \ln x \, dx = -\frac{x^2(2 \ln x - 1)}{8}.$$

The general solution is therefore

$$y(x) = c_1 x + \frac{c_2}{x} + \frac{1}{4} x (\ln x)^2 - \frac{1}{4} x \ln x.$$

It appears as if one term from u_2 has been omitted from $y(x)$. This is simply because $(x^2/8)(1/x) = x/8$ and such a term adds nothing new in view of the presence of $c_1 x$ in the complementary function — it involves only a redefinition of the integration constant c_1.   □

4.5　REDUCTION OF ORDER

Reduction of order is a procedure which allows the order of a differential equation to be reduced by 1 if a solution of the homogeneous equation is known. The result is not surprising since the knowledge of a solution to the equation is tantamount to reducing the degree of freedom of the equation by one. The method is illustrated with respect to second order differential equations. Let $y(x) = y_1(x)$ be a solution of the second order equation

$$y''(x) + p(x)y'(x) + q(x)y(x) = 0. \tag{4.20}$$

Substitute

$$y_2(x) = u(x)y_1(x) \tag{4.21}$$

into (4.20) to obtain

$$y_1 u'' + (2y_1' + py_1)u' + (y_1'' + py_1' + qy_1)u = 0.$$

The third term is seen to be zero since y_1 is a solution of (4.20). Now substitute $w(x) = u'(x)$ (and with it $w' = u''$) to get[1]

$$y_1 w' + (2y_1' + py_1)w = 0$$

which is a first order separable equation for $w(x)$. Its solution is

$$w(x) = e^{-\int (2y_1'/y_1 + p)\, dx} = e^{-2\ln y_1}\, e^{-\int p\, dx} = \frac{1}{y_1^2} e^{-\int p\, dx}$$

so that

$$u(x) = \int^x w(s)ds \qquad \longrightarrow \qquad y_2(x) = u(x)y_1(x)$$

and the general solution is

$$y(x) = c_1 y_1(x) + c_2 y_2(x) = c_1 y_1(x) + c_2 u(x)y_1(x).$$

Remarks 4.2　Note that this procedure can also be applied to *inhomogeneous* equations.　　　　　　　　　　　　　　　　　　　　　　　　△

Example 4.6　Show that $y_1(x) = e^x$ is a solution of the differential equation $xy'' - (x+1)y' + y = 0$ and find the general solution.

---

[1] We have already encountered this trick in Section 2.4.

Solution 4.6   Direct substitution confirms e^x as a solution. Next, rewrite the equation in standard form

$$y'' - \frac{x+1}{x} y' + \frac{y}{x} = 0 \,.$$

Thus

$$w(x) = \frac{e^{\int [(x+1)/x]\,dx}}{e^{2x}} = \frac{e^{(x+\ln x)}}{e^{2x}} = xe^{-x}$$

and so

$$y_2(x) = y_1(x)u(x) = e^x \int^x se^{-s}ds = e^x(-xe^{-x} - e^{-x}) = -x - 1 \,.$$

The general solution is finally

$$y = c_1 e^x + c_2(x+1) \,. \qquad \square$$

Example 4.7   Find the general solution of the differential equation

$$x^3 \frac{d^2 y}{dx^2} + x\frac{dy}{dx} - y = 0$$

by observing that $y_1(x) = x$ is one of its solutions.

Solution 4.7   Since $y_1(x) = x$ satisfies the homogeneous equation we write $y_2(x) = xv(x)$ where v is the new dependent variable. The differential equation for $v(x)$ is

$$x^3 \left(x\frac{d^2 v}{dx^2} + 2\frac{dv}{dx} \right) + x \left(x\frac{dv}{dx} + v \right) - xv = 0$$

which may be simplified to

$$x^2 \frac{d^2 v}{dx^2} + (2x+1)\frac{dv}{dx} = 0 \qquad \longrightarrow \qquad \frac{dz}{dx} + \left(\frac{2}{x} + \frac{1}{x^2} \right) z = 0$$

where $z = dv/dx$. This equation has integrating factor $x^2 e^{(-1/x)}$ and may be integrated to get

$$z = \frac{C}{x^2} e^{(1/x)} \,.$$

A further integration now yields

$$v(x) = \int^x z(s)\,ds = -Ce^{(1/x)} + B \qquad \longrightarrow \qquad y_2(x) = Bx - Cx\,e^{(1/x)} \,.$$

The general solution is therefore

$$y(x) = Dx + Ex\,e^{(1/x)} \,. \qquad \square$$

4.6 EULER'S DIFFERENTIAL EQUATION

The differential equation

$$ax^2 y''(x) + bxy'(x) + cy(x) = 0 \tag{4.22}$$

where a, b and c are (real–valued) constants, is called **Euler's differential equation** (or a *differential equation of Euler's type*). It has at least one solution of type $y(x) = x^\lambda$. Upon substitution and collection of terms, the equation is seen to be fulfilled provided λ satisfies the quadratic equation

$$a\lambda(\lambda - 1) + b\lambda + c = 0. \tag{4.23}$$

There are three different cases to be considered.

Case 1: The roots λ_1 and λ_2 of (4.23) are both real–valued and distinct from each other. The general solution is

$$y(x) = c_1 x^{\lambda_1} + c_2 x^{\lambda_2}. \tag{4.24}$$

Case 2: The roots form a complex conjugate pair $\lambda_{1,2} = \alpha \pm i\beta$. Similarly,

$$y(x) = c_1 x^{\alpha + i\beta} + c_2 x^{\alpha + i\beta}. \tag{4.25}$$

This expression can be put into more convenient form

$$
\begin{aligned}
y(x) &= x^\alpha \left(c_1 x^{i\beta} + c_2 x^{-i\beta} \right) = x^\alpha \left(c_1 e^{i\beta \ln x} + c_2 e^{-i\beta \ln x} \right) \\
&= x^\alpha \left[d_1 \cos(\beta \ln x) + d_2 \sin(\beta \ln x) \right].
\end{aligned} \tag{4.26}
$$

Case 3: There is a double root $\lambda_{1,2} = (a - b)/2a$. Therefore this procedure gives only one solution. A second solution can be found by using variation of constants. Substitute $y(x) = A(x)x^\lambda$ with $\lambda = (a - b)/2a$ into (4.22). After some simple calculations, it is found that

$$A''(x) + \frac{A'(x)}{x} = 0 \quad \longrightarrow \quad A'(x) = \frac{1}{x} \quad \longrightarrow \quad A(x) = \ln x.$$

The general solution in this instance is

$$y(x) = x^\lambda (c_1 + c_2 \ln x). \tag{4.27}$$

Remarks 4.3 Note the correspondence between Euler's differential equation $ax^2 y''(x) + bxy'(x) + cy(x) = 0$ and the equation $\ddot{y}(t) + \gamma \dot{y}(t) + \delta y(t) = 0$ with constant coefficients. One can be transformed into the other by changing independent variable from x to t using $x = e^t$. This connection between the two equations will be explored further in Tutorial example T 4.17. △

Example 4.8   Find the general solution of the second order differential equation

$$xy''(x) - \beta y'(x) + \frac{\beta}{x} y(x) = 0, \qquad \beta > 0.$$

Solution 4.8   The differential equation is clearly of Euler's type. Thus, substitution of $y(x) = x^\lambda$ leads to the relation

$$\lambda^2 - (\beta + 1)\lambda + \beta = 0$$

which has the roots $\lambda_1 = 1$ and $\lambda_2 = \beta$. Consequently, the general solution is given by

$$y(x) = \begin{cases} Ax + Bx^\beta & \beta \neq 1 \\ (A + B\ln x)\,x & \beta = 1. \end{cases} \qquad \Box$$

4.7   EXACT EQUATIONS

Suppose that $y(x)$ satisfies the second order linear differential equation

$$a(x)\frac{d^2y}{dx^2} + b(x)\frac{dy}{dx} + c(x)y = f(x), \tag{4.28}$$

then it can be verified that this equation may be rewritten as

$$\frac{d}{dx}\left[ay' + (b - a')y\right] + (a'' - b' + c)y = f(x). \tag{4.29}$$

Equation (4.28) is said to be an **exact** equation if and only if

$$a''(x) - b'(x) + c(x) = 0. \tag{4.30}$$

In this event, (4.29) can be integrated once to obtain

$$ay' + (b - a')y = \int^x f(s)\,ds \tag{4.31}$$

which is a differential equation of first order that can be solved accordingly.

Example 4.9   Find the general solution of the inhomogeneous equation

$$x^2 y'' + 3xy' + y = \ln x.$$

Solution 4.9   Because $(x^2)'' - (3x)' + 1 = 0$, the equation is exact and can be rewritten as

$$\frac{d}{dx}\left(x^2 y' + xy\right) = \ln x.$$

T 4.14   Solve the initial value problem

$$x^2 y'' + 4xy' + 2y = 0, \qquad y(a) = a, \quad y'(a) = -1.$$

T 4.15   Solve the initial value problem

$$x^2 y'' - xy' - 3y = x^2, \qquad y(1) = y'(1) = 0.$$

T 4.16   Solve the initial value problem

$$x^2 y'' + ay = 0, \quad (a > 1/4), \qquad y(1) = 0, \quad y'(1) = 1.$$

[*Hint*: remove the imaginary powers of x in the solution by transforming to exponential functions.]

T 4.17   Show that Euler's differential equation of order n

$$x^n y^{(n)} + a_1 x^{n-1} y^{(n-1)} + \dots + a_{n-1} xy' + a_n y = f(x)$$

can be transformed into a linear differential equation with constant coefficients by defining a new independent variable t such that $x = e^t$.

T 4.18   Show that the function $y_1(x) = x^2$ is a solution of the homogeneous version of the differential equation

$$x^2 y'' - 3xy' + 4y = x^2 \ln x, \qquad x > 0.$$

Obtain the general solution of this inhomogeneous equation.

T 4.19   Show that the differential equation

$$xy'' + 2(1 + x)(y' + y) = 0$$

is not exact but can be made exact by multiplying it by any solution of the differential equation $\phi'' - 2\phi' + 2\phi = 0$. Hence determine the solution of the original differential equation satisfying the initial condition $y \to 1$ as $x \to 0$.

T 4.20   Find the general solution of the differential equation

$$x^2 y'' + 4x(1 + x)y' + (8x + 2)y = 0$$

and the particular solution satisfying $y(1) = 0$ and $y'(1) = 1$.

T 4.21   By using the substitution

$$u(x) = e^{-\int p(x)y(x)\,dx}$$

show that the *Riccati* equation

$$y'(x) = p(x)y^2(x) + q(x)y(x) + r(x)$$

can be transformed into the linear differential equation of second order

$$u'' - \left(q + \frac{p'}{p}\right)u' + rpu = 0$$

and use this information to obtain the general solution of

$$y' = -e^x y^2 - y + 4e^{-x}.$$

T 4.22   A self–gravitating star in the form of a sphere of radius a is composed of a compressible fluid whose pressure p and density ρ are connected by the formula $p = k\rho^2$ where k is a positive constant. The acceleration $g(r)$ due to gravity, pressure $p(r)$ and density $\rho(r)$ at radius r are also connected by equations

$$\frac{dp}{dr} = -g(r)\rho(r)\,, \qquad r^2 g(r) = 4\pi G \int_0^r s^2 \rho(s)\,ds$$

(G is the gravitational constant), expressing balance of momentum and the law of gravity respectively. Deduce that ρ satisfies the differential equation

$$r\rho'' + 2\rho' + \alpha^2 r\rho = 0\,, \qquad \alpha^2 = \frac{2\pi G}{k}$$

where the prime denotes differentiation with respect to r. Determine $\rho(r)$ in terms of $\rho(0)$, the density at the stellar core. Explain why this stellar model predicts a maximum size for stars.

5

Oscillatory Motion

5.1 DAMPED OSCILLATORY MOTION

It was already established in Section 3.1 that real damped systems are described by the generic differential equation

$$\ddot{y} + 2\alpha\dot{y} + \omega_0^2 y = f(t)$$

where ω_0 is the **undamped frequency** of the system, α is a positive **damping parameter** and $f(t)$ is an **externally applied force** acting on the system. In mechanical situations, $\alpha = b/2m$ and $\omega_0^2 = k/m$ while in the context of electrical circuit theory, $\alpha = R/2L$ and $\omega_0^2 = 1/LC$, where the constants have the meaning denoted in Section 3.1.

The auxiliary equation is $\lambda^2 + 2\alpha\lambda + \omega_0^2 = 0$ with roots

$$\lambda_{1,2} = -\alpha \pm \sqrt{\alpha^2 - \omega_0^2}. \tag{5.1}$$

The solutions of this quadratic are classified into three separate physical regimes depending on the relative values of α and ω_0.

Heavy damping: $\alpha > \omega_0$

Here the auxiliary equation has two real–valued negative roots

$$\lambda_1 = -\alpha - \sqrt{\alpha^2 - \omega_0^2}, \qquad \lambda_2 = -\alpha + \sqrt{\alpha^2 - \omega_0^2} \tag{5.2}$$

and the complementary function is

$$y_c(t) = Ae^{\lambda_1 t} + Be^{\lambda_2 t}. \tag{5.3}$$

This possibility is often called the *overdamped* case since no oscillations occur and the solution decays exponentially to zero as $t \to \infty$. If, for example,

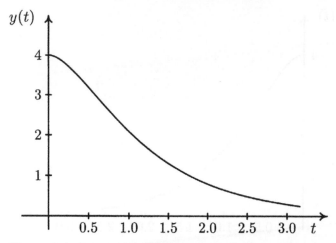

Figure 5.1: Damped oscillatory motion. Heavy damping.

this equation modelled an automobile suspension system, the ride would be hard and uncomfortable. The equation

$$\ddot{y} + 4\dot{y} + 3y = 0, \qquad y(0) = 4, \quad \dot{y}(0) = 0$$

provides an example of an overdamped system. An illustration of the solution $y(t) = 6e^{-t} - 2e^{-3t}$ is provided in Figure 5.1.

Critical damping: $\alpha = \omega_0$

Here the auxiliary equation has two equal real solutions $\lambda_{1,2} = -\alpha$ and the related complementary function is

$$y_c(t) = (A + Bt)e^{-\alpha t}. \tag{5.4}$$

This possibility is usually called *critical damping* and is often the desired configuration for practical application since it represents the weakest damping before oscillatory behaviour becomes possible. As in the previous case, this solution decays to zero as $t \to \infty$. An example is provided by the initial value problem

$$\ddot{y} + 4\dot{y} + 4y = 0, \qquad y(0) = 4, \quad \dot{y}(0) = 1$$

which has the critically damped solution $y(t) = (4 + 8t)e^{-2t}$. The graph of this function is shown in Figure 5.2.

Light damping: $\alpha < \omega_0$

Here the auxiliary equation has two complex conjugate roots

$$\lambda_{1,2} = -\alpha \pm i\omega_n, \qquad \omega_n = \sqrt{\omega_0^2 - \alpha^2} \tag{5.5}$$

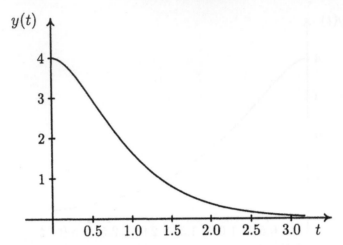

Figure 5.2: Damped oscillatory motion. Critical damping.

where ω_n is called the **natural frequency** of the system. The complementary function is

$$y_c(t) = e^{-\alpha t} \left(A \cos \omega_n t + B \sin \omega_n t \right) \tag{5.6}$$

which can be recast in the form

$$y_c(t) = C e^{-\alpha t} \sin \left(\omega_n t + \phi \right) \tag{5.7}$$

where the integration constants in (5.6) and (5.7) are related through

$$C = \sqrt{A^2 + B^2}, \qquad \tan \phi = A/B.$$

As previously, this solution decays to zero as $t \to \infty$ but the decay process is now oscillatory in nature. The quantity $T = 2\pi/\omega_n$ is the **period** of the damped oscillation. Furthermore, as the absolute value of the trigonometric function in (5.7) does not exceed 1, the curves $C e^{-\alpha t}$ and $-C e^{-\alpha t}$ provide an envelope for the damped oscillation.

The initial value problem described by

$$\ddot{y} + 4\dot{y} + 68y = 0, \qquad y(0) = 4, \quad \dot{y}(0) = 0$$

has solution $y = e^{-2t}(\sin 8t + 4 \cos 8t)$ and is *lightly damped*. The graph of this function is shown in Figure 5.3.

5.2   FORCED OSCILLATIONS

Probably the most important application of second order equations with constant coefficients arises when the applied force is given by

$$f(t) = F \cos \left(pt \right) \tag{5.8}$$

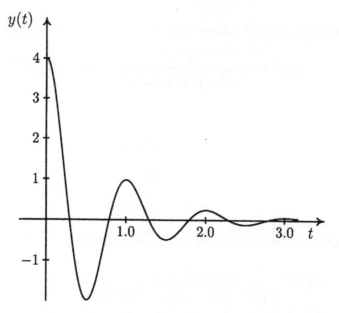

Figure 5.3: Damped oscillatory motion. Light damping.

in which F and p are real–valued constants. This is well known as the forced oscillation problem. Here it is required to solve the differential equation

$$\ddot{y} + 2\alpha\dot{y} + \omega_0^2 y = F\cos(pt). \tag{5.9}$$

Fourier series methods allow many realistic waveforms of periodic shape and finite energy to be represented as a sum of cosine and/or sine components. The analysis here may be regarded as the first step in the treatment of a general periodic right–hand side. Since $F\cos(pt) = F\,\mathrm{Re}(e^{ipt})$, then the required particular integral can be constructed from the real part of $z = Ae^{ipt}$ by using the method developed in Section 4.3. Clearly,

$$\ddot{z} + 2\alpha\dot{z} + \omega_0^2 z = Fe^{ipt} \longrightarrow A\left[(ip)^2 + 2\alpha(ip) + \omega_0^2\right]e^{ipt} = Fe^{ipt}.$$

Hence if A is chosen to satisfy

$$A = \frac{F}{(\omega_0^2 - p^2) + 2i\alpha p} = \frac{F}{\sqrt{(\omega_0^2 - p^2)^2 + 4\alpha^2 p^2}}\,e^{-i\phi}$$

$$\tan\phi = \frac{2\alpha p}{\omega_0^2 - p^2},$$

then the particular integral pertaining to Fe^{ipt} is

$$Ae^{ipt} = \frac{F}{\sqrt{(\omega_0^2 - p^2)^2 + 4\alpha^2 p^2}}\,e^{i(pt-\phi)},$$

and by taking real parts, it follows that the particular integral for $F\cos pt$ is

$$y_p(t) = \frac{F}{\sqrt{(\omega_0^2 - p^2)^2 + 4\alpha^2 p^2}} \cos(pt - \phi). \qquad (5.10)$$

Since

$$
\begin{aligned}
(\omega_0^2 - p^2)^2 + 4\alpha^2 p^2 &= p^4 + (4\alpha^2 - 2\omega_0^2)p^2 + \omega_0^4 \\
&= \left[p^2 - (\omega_0^2 - 2\alpha^2)\right]^2 + 4\alpha^2(\omega_0^2 - \alpha^2) \\
&= \left[p^2 - (\omega_0^2 - 2\alpha^2)\right]^2 + 4\alpha^2\omega_n^2
\end{aligned}
$$

then the particular integral (5.10) can be rewritten as

$$y_p(t) = \frac{F}{\sqrt{[p^2 - (\omega_0^2 - 2\alpha^2)]^2 + 4\alpha^2\omega_n^2}} \cos(pt - \phi). \qquad (5.11)$$

Thus for a given amplitude F of driving force, the amplitude of $y_p(t)$ is maximum at frequency $p = \omega_r$ where

$$\omega_r = \sqrt{\omega_0^2 - 2\alpha^2}. \qquad (5.12)$$

The frequency ω_r is called the **resonance frequency** of the system. Evidently, $\omega_r \le \omega_n \le \omega_0$, (if the damping parameter α vanishes, all three frequencies are the same) and generally ω_r, ω_n and ω_0 satisfy

$$\omega_n^2 = \frac{1}{2}(\omega_0^2 + \omega_r^2).$$

Provided the damping parameter α is positive, the complementary function decays to zero, and in practice this decay is often rapid. Therefore the response of the system to the periodic driving force is well described by expression (5.11) for large t.

The Q–factor

Let $a(p)$ be the amplitude of the damped oscillation induced by a driving force of amplitude F and frequency p, then from (5.11)

$$a(p) = \frac{F}{\sqrt{(p^2 - \omega_r^2)^2 + 4\alpha^2\omega_n^2}}. \qquad (5.13)$$

The **Q–factor** or *quality factor* of an oscillator is defined to be the ratio $a(p)/a(0)$. This is a non–dimensional number corresponding to the ratio of the amplitude of oscillations induced by a periodic driving force at frequency p to the deformation induced by a force of the same amplitude but applied statically. Thus

$$Q = \frac{a(p)}{a(0)} = \frac{\omega_0^2}{\sqrt{(p^2 - \omega_r^2)^2 + 4\alpha^2\omega_n^2}}. \tag{5.14}$$

Figure 5.4 gives the general shape of Q for frequencies ranging from $p = 0$ (static load) to frequencies in excess of the resonant frequency ω_r. Under all circumstances, $1 \le Q \le Q_{\max}$ where $Q_{\max} = \omega_0^2/(2\alpha\omega_n)$ is the maximum value of Q occurring at the resonant frequency ω_r. For lightly damped oscillators, α is small so that $\omega_0 \approx \omega_r \approx \omega_n$ and $Q_{\max} \approx \omega_0/(2\alpha)$. In particular, small values of α lead to sharp resonance and a large value for $Q_{\max}$.

The quality of the resonance at $p = \omega_r$ is often quantified in terms of the size of the band of frequencies about ω_r for which $Q \ge Q_{\max}/\sqrt{2}$. After some algebra, it may be seen that this condition is satisfied provided

$$\omega_r - \frac{\alpha\omega_n}{\omega_r} \approx \sqrt{-2\alpha\omega_n + \omega_r^2} \le p \le \sqrt{2\alpha\omega_n + \omega_r^2} \approx \omega_r + \frac{\alpha\omega_n}{\omega_r} \tag{5.15}$$

in which the approximations are obtained by expanding the square roots in this inequality as power series in α. This is a sensible strategy since α is small whenever resonance is sharp. Therefore $Q \ge Q_{\max}/\sqrt{2}$ within a band of frequencies of approximate width $2\alpha\omega_n/\omega_r$ centred on ω_r. In view of the definition of $Q_{\max}$, the width of this band is approximately $\Delta\omega \approx \omega_n/Q_{\max}$. Thus $Q_{\max}$ measures tuning quality in terms of the ability of the oscillator to isolate frequencies close to resonance and discard those that are not.

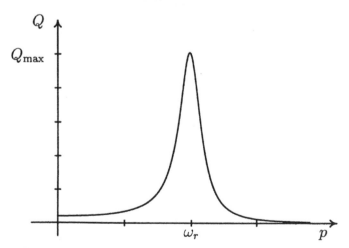

Figure 5.4: General form of a frequency response curve.

5.3   TUTORIAL EXAMPLES 5

T 5.1   The deviation of the glucose concentration $g(t)$ from its base level in a human body can be modelled by the second order differential equation

$$\ddot{g} + 2\alpha\dot{g} + \omega_0^2 g = 0, \qquad \alpha > 0,$$

with the initial conditions $g(0) = 0$, $\dot{g}(0) = \beta$ where β is the initial (unknown) glucose uptake rate. Akerman proposed a diabetes test in which fasted patients (glucose level at base) were required to ingest a large dose of glucose and thereafter its level in their blood was monitored hourly. Non-diabetic patients had values of ω_0 corresponding to undamped oscillations with maximum period 4 hours whereas varying degrees of diabetes were attributed to patients with periods exceeding 4 hours.

A patient arrives at a clinic with a fasted glucose level of 0.70 mg/cm^3 of blood. Levels of 1.00, 0.55 and 0.75 mg/cm^3 were measured 1 hour, 2 hours and 3 hours respectively after administration of a strong glucose drink. Classify this patient.

T 5.2   A damped oscillatory system has period 1 second and the oscillations are damped to half amplitude in 40 complete oscillations. Describe the qualitative effects on the period and the number of oscillations to half the amplitude brought about by slightly increasing or decreasing the damping and slightly increasing or decreasing the restoring effect. Assume only one parameter is changed in each instance.

Extract the second order differential equation with the given period and damping characteristics.

6

Laplace Transform

6.1   GENERAL INTEGRAL TRANSFORMS

The Laplace Transform provides an efficient method for solving initial value problems for ordinary (and also partial) differential equations by converting the differential equation into an algebraic equation. The solution of the differential equation can then be obtained from the solution of the algebraic problem when the transformation is inverted.

A general **integral transform** is an expression of the type

$$\overline{f}(s) = \int_{\alpha}^{\beta} K(s,t) f(t) \, dt \tag{6.1}$$

which transforms a given function $f(t)$ into another function $\overline{f}(s)$. The function $K(s,t)$ is called the **kernel** of the transform.

6.2   DEFINITION OF THE LAPLACE TRANSFORM

The Laplace transform is the special case of (6.1) when $\alpha = 0$, $\beta = \infty$ and $K(s,t) = e^{-st}$. Therefore the **Laplace transform** of $f(t)$ is defined by

$$\overline{f}(s) = L[f(t) : t \to s] = \int_{0}^{\infty} e^{-st} f(t) \, dt \,. \tag{6.2}$$

Note that the integral in (6.2) is *improper*. Here $f(t)$ will be regarded as a real–valued function defined for $t \geq 0$. Reversing the transformation procedure leads to the **inverse Laplace transform**. If $L[f(t) : t \to s]$ then

$$L^{-1}\left[\overline{f}(s) : s \to t\right] = f(t) \,. \tag{6.3}$$

In other words, the original function $f(t)$ is the inverse Laplace transform of $\overline{f}(s) = L[f(t) : t \to s]$.

Two important and immediate questions are: (1) what functions have a Laplace transform? and (2) can two functions $f(t)$, $g(t)$ have the same Laplace transform (i.e., is the inverse Laplace transform unique)? The answers are provided by the following statement: The existence and uniqueness of the Laplace transform of a function $f(t)$ is guaranteed if there exist real K, M and a such that

1. $f(t)$ is piecewise continuous for $t \geq 0$.

2. $|f(t)| \leq Ke^{at}$ for $t \geq M$.

Example 6.1   Find the Laplace transform of the function $f(t) = 1$.

Solution 6.1

$$L[1 : t \to s] = \int_0^\infty e^{-st}\, dt = \left[\frac{e^{-st}}{-s}\right]_0^\infty = \frac{1}{s}.$$

Note that the validity of the result is limited to $s > 0$ because otherwise the integral does not exist.   □

Example 6.2   Find the Laplace transform of the function $f(t) = t$.

Solution 6.2

$$L[t : t \to s] = \int_0^\infty te^{-st}\, dt = \left[\frac{te^{-st}}{-s}\right]_0^\infty - \int_0^\infty \frac{e^{-st}}{(-s)}\, dt$$

$$= \frac{1}{s}\left[\frac{e^{-st}}{-s}\right]_0^\infty = \frac{1}{s^2}, \qquad (s > 0).\qquad\qquad □$$

Example 6.3   Find the Laplace transform of the function $f(t) = e^{\alpha t}$.

Solution 6.3

$$L\left[e^{\alpha t} : t \to s\right] = \int_0^\infty e^{\alpha t}e^{-st}\, dt = \left[\frac{e^{(\alpha - s)t}}{\alpha - s}\right]_0^\infty = \frac{1}{s - \alpha}, \qquad (s > \alpha). \quad □$$

Example 6.4   Find the Laplace transform of the functions $f(t) = \sin at$ and $f(t) = \cos at$.

Solution 6.4

$$L[\cos at + i\sin at : t \to s] \;=\; L\left[e^{iat} : t \to s\right] = \frac{1}{s - ia}$$

$$= \frac{s + ia}{s^2 + a^2}, \qquad (s > 0).$$

Separating real and imaginary parts yields

$$L[\cos at : t \to s] = \frac{s}{s^2 + a^2}\,, \qquad L[\sin at : t \to s] = \frac{a}{s^2 + a^2}\,.$$

Alternatively, the Laplace transforms of $\cos at$ and $\sin at$ may be computed directly using integration by parts.   □

Selected functions and their Laplace transforms are listed in Table 6.1.

6.3 SOME GENERAL PROPERTIES

Some important general properties of the Laplace transform that follow directly from its definition as a quadrature are now listed.

1. **Linearity.** If $\overline{f}(s) = L[f(t) : t \to s]$ and $\overline{g}(s) = L[g(t) : t \to s]$ exist then $L[af(t) + bg(t) : t \to s]$ exists for all constants a and b and

$$L[af(t) + bg(t) : t \to s] = a\,\overline{f}(s) + b\,\overline{g}(s)\,. \tag{6.4}$$

2. **First shifting property.** Suppose $\overline{f}(s) = L[f(t) : t \to s]$ exists and that a is a constant then $L[e^{at}f(t) : t \to s]$ exists and

$$L\left[e^{at}f(t) : t \to s\right] = \overline{f}(s - a)\,, \tag{6.5}$$

or, conversely,

$$L^{-1}\left[\overline{f}(s - a) : s \to t\right] = e^{at}f(t)\,. \tag{6.6}$$

6.4 CONVOLUTION

Convolution integrals provide an important way to extend the usefulness of Laplace transforms by providing a means by which many more exact transforms can be calculated or expressed as quadratures. Let

$$\overline{f}(s) = L[f(t) : t \to s]\,, \qquad \overline{g}(s) = L[g(t) : t \to s]$$

be the Laplace transforms of $f(t)$ and $g(t)$ respectively. The linearity of the Laplace transform operator guarantees that

$$\begin{aligned} L^{-1}\left[\overline{f}(s) + \overline{g}(s) : s \to t\right] &= L^{-1}\left[\overline{f}(s) : s \to t\right] + L^{-1}[g(s) : s \to t] \\ &= f(t) + g(t)\,. \end{aligned} \tag{6.7}$$

However,

$$L^{-1}\left[\overline{f}(s)\overline{g}(s) : s \to t\right] \neq f(t)\,g(t)\,, \tag{6.8}$$

$f(t) = L^{-1}\left[\overline{f}(s) : s \to t\right]$	$\overline{f}(s) = L\left[f(t) : t \to s\right]$	
1.	1	$\dfrac{1}{s}$	$\operatorname{Re}(s) > 0$
2.	e^{at}	$\dfrac{1}{s-a}$	$\operatorname{Re}(s) > a$
3.	$t^n\,(n : \text{positive integer})$	$\dfrac{n!}{s^{n+1}}$	$\operatorname{Re}(s) > 0$
4.	$t^p\,(p > -1)$	$\dfrac{\Gamma(p+1)}{s^{p+1}}$	$\operatorname{Re}(s) > 0$
5.	$\cos at$	$\dfrac{s}{s^2 + a^2}$	$\operatorname{Re}(s) > 0$
6.	$\sin at$	$\dfrac{a}{s^2 + a^2}$	$\operatorname{Re}(s) > 0$
7.	$\cosh at$	$\dfrac{s}{s^2 - a^2}$	$\operatorname{Re}(s) > |a|$
8.	$\sinh at$	$\dfrac{a}{s^2 - a^2}$	$\operatorname{Re}(s) > |a|$
9.	$e^{at}\cos bt$	$\dfrac{s-a}{(s-a)^2 + b^2}$	$\operatorname{Re}(s) > a$
10.	$e^{at}\sin bt$	$\dfrac{b}{(s-a)^2 + b^2}$	$\operatorname{Re}(s) > a$
11.	$t^n e^{at}\,(n : \text{positive integer})$	$\dfrac{n!}{(s-a)^{n+1}}$	$\operatorname{Re}(s) > a$

Table 6.1: Laplace transforms of selected functions ($\Gamma(x)$ is the Gamma function defined in Tutorial example T 6.3).

so that the inverse Laplace transform of a product does *not* equal the product of the inverse Laplace transforms of the factors. For example,

$$\overline{f}(s) = \frac{1}{s}, \quad \overline{g}(s) = \frac{1}{s^2} \quad \longrightarrow \quad f(t) = 1, \quad g(t) = t.$$

Thus $f(t)g(t) = t$ while

$$L^{-1}\left[\overline{f}(s)\overline{g}(s) : s \to t\right] = L^{-1}\left[1/s^3 : s \to t\right] = t^2/2,$$

which does not equal $f(t)g(t) = t$.

Given two functions $f(t)$ and $g(t)$, the **convolution** of f and g, denoted $f * g$, is defined by the integral

$$(f * g)(t) = \int_0^t f(t-u)g(u)\, du \qquad (6.9)$$

whenever this integral exists. A simple change of variable shows that the convolution is symmetric, that is, $(f * g)(t) = (g * f)(t)$. Furthermore, by change of order in double integration, it can also be shown that

$$L[(f * g)(t) : t \to s] = \overline{f}(s)\,\overline{g}(s), \qquad (6.10)$$

or, conversely,

$$L^{-1}\left[\overline{f}(s)\,\overline{g}(s) : s \to t\right] = (f * g)(t). \qquad (6.11)$$

To appreciate results (6.10) and (6.11), we begin by recognizing that

$$L[(f * g)(t) : t \to s] = \int_0^\infty \left(\int_0^t f(t-u)g(u)\, du\right) e^{-st}\, dt. \qquad (6.12)$$

The region of double integration in (6.12) is illustrated in Figure 6.1.

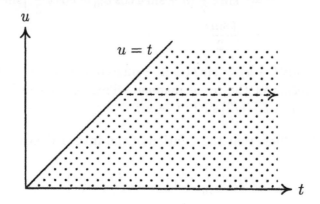

Figure 6.1: The region of double integration showing the new order of integration in which t is integrated first from $t = u$ to $t = \infty$ followed by integration of u from $u = 0$ to $u = \infty$.

On changing the order of integration in the right–hand side of (6.12), the Laplace transform of $f * g$ becomes

$$
\begin{aligned}
L[(f * g)(t) : t \to s] &= \int_0^\infty g(u) \left(\int_u^\infty f(t-u)\, e^{-st}\, dt\right) du \\
&= \int_0^\infty g(u) \left(\int_0^\infty f(w)\, e^{-s(w+u)}\, dw\right) du
\end{aligned}
$$

$$= \int_0^\infty g(u) \, e^{-su} \left(\int_0^\infty f(w) \, e^{-sw} \, dw \right) du$$

$$= \bar{f}(s) \, \bar{g}(s)$$

which establishes (6.10).

Example 6.5 Compute $L^{-1}\left[s/(s^2+1)^2 : s \to t \right]$.

Solution 6.5 The convolution property (6.11) asserts that

$$L^{-1}\left[\frac{s}{(s^2+1)^2} : s \to t \right] = L^{-1}\left[\frac{s}{s^2+1} \frac{1}{s^2+1} : s \to t \right] = (\sin * \cos)(t)$$

where it has been recognized that $L^{-1}\left[s/(s^2+1) : s \to t \right] = \cos t$ and also that $L^{-1}\left[1/(s^2+1) : s \to t \right] = \sin t$. Now,

$$
\begin{aligned}
(\sin * \cos)(t) &= \int_0^t \sin(t-u) \cos u \, du \\
&= \sin t \int_0^t \cos^2 u \, du - \cos t \int_0^t \sin u \cos u \, du \\
&= \sin t \, \frac{1}{2} \left[u + \sin u \cos u \right]_0^t - \cos t \, \frac{1}{2} \left[\sin^2 u \right]_0^t \\
&= \frac{t \sin t}{2}
\end{aligned}
$$

where we used $\sin(t-u) = \sin t \cos u - \sin u \cos t$ and then integrated the trigonometric terms in the usual manner. This result can be checked by direct calculation of $L\left[(t/2) \sin t : t \to s \right]$. □

Example 6.6 Show that the solution of the **integral equation**

$$f(t) + \int_0^t e^u \, f(t-u) \, du = g(t)$$

can be written in the form

$$f(t) = g(t) - \int_0^t g(u) \, du .$$

Solution 6.6 Let $\bar{f}(s)$ be the Laplace transform of $f(t)$. Taking Laplace transforms of the integral equation and using the fact that the Laplace transform of a convolution is the product of the Laplace transforms of the component functions, leads to

$$\bar{f} + \bar{f} L \left[e^t : t \to s \right] = \bar{g} \qquad \longrightarrow \qquad \bar{f} = \bar{g} - \frac{\bar{g}}{s} .$$

Therefore

$$f(t) = g(t) - L^{-1}\left[\frac{\overline{g}(s)}{s} : s \to t\right] = g(t) - \int_0^t g(x)\,dx.$$

The inversion of the latter Laplace transform is done by interpreting it as a convolution of $g(x)$ with the constant function 1 because the Laplace transform of 1 is $1/s$. □

6.5 APPLICATION TO INITIAL VALUE PROBLEMS

Laplace transforms are primarily used in the solution of *initial value problems*. Suppose that f has a Laplace transform and that $\dot{f}$ is piecewise continuous in the interval $(0, \infty)$ (note that the result can be generalized to functions f for which f and $\dot{f}$ have jump discontinuities). Thus

$$
\begin{aligned}
L\left[\dot{f}(t) : t \to s\right] &= \int_0^\infty \dot{f}(t)e^{-st}\,dt \\
&= \left[f(t)e^{-st}\right]_0^\infty + s\int_0^\infty f(t)e^{-st}\,dt \\
&= s\overline{f}(s) - f(0) \qquad\qquad (6.13)
\end{aligned}
$$

and similarly

$$
\begin{aligned}
L\left[\ddot{f}(t) : t \to s\right] &= \int_0^\infty \ddot{f}(t)e^{-st}\,dt = \left[\dot{f}(t)e^{-st}\right]_0^\infty + s\int_0^\infty \dot{f}(t)e^{-st}\,dt \\
&= -\dot{f}(0) + sL\left[\dot{f}(t) : t \to s\right] \\
&= s^2\overline{f}(s) - sf(0) - \dot{f}(0). \qquad\qquad (6.14)
\end{aligned}
$$

Both (6.13) and (6.14) are valid for $s > 0$.

Example 6.7 Find the solution of the initial value problem

$$\ddot{y} - 4y = 0, \qquad y(0) = 1, \quad \dot{y}(0) = 2.$$

Solution 6.7 Let $\overline{y}(s) = L[y(t) : t \to s]$. When the Laplace transform operation is applied to the differential equation and result (6.14) is used, $\overline{y}(s)$ is seen to satisfy

$$s^2\overline{y}(s) - sy(0) - \dot{y}(0) - 4\overline{y}(s) = 0$$

which can be solved algebraically for $\overline{y}(s)$ to give

$$\overline{y}(s) = \frac{sy(0) + \dot{y}(0)}{s^2 - 4} = \frac{s + 2}{s^2 - 4} = \frac{1}{s - 2}.$$

The solution can now be extracted from Table 6.1 and is

$$Q(t) = \frac{1}{200}(4\cos t + 3\sin t - 4e^{-3t} - 15te^{-3t})$$

by using the first shifting property (6.5) or formula (6.16) for the fourth term. Once $Q(t)$ is known, the current $I(t)$ follows immediately from $I(t) = \dot{Q}(t)$ and is given by

$$I(t) = \dot{Q}(t) = \frac{1}{200}(-4\sin t + 3\cos t - 3e^{-3t} + 45te^{-3t}).$$

Furthermore, the long–term behaviour of the solution is described by the result

$$\lim_{t\to\infty}\left[Q(t) - \frac{1}{200}(4\cos t + 3\sin t)\right] = 0.$$

This means that the solution for large time is dominated by the inhomogeneous term in the differential equation, while the contributions from the complementary function are damped out. In this example the forcing function $U(t) = 250\cos t$ was a continuous function of time t. This is not always the case; in many applications, the forcing function is a step function or an impulse function.

We note that the present problem could also have been solved as an initial value problem for $I(t)$ instead. Differentiation of (6.17) yields

$$L\ddot{I}(t) + R\dot{I}(t) + \frac{I}{C} = \dot{U}(t) \tag{6.18}$$

where we have again used $\dot{Q} = I$. The initial conditions are

$$I(0) = \dot{Q}(0), \qquad \dot{I}(0) = \frac{1}{L}\left[U(0) - R\dot{Q}(0) - \frac{Q(0)}{C}\right]$$

where the condition for $\dot{I}(0)$ follows from evaluation of the differential equation (6.17) at $t = 0$.   □

6.6   THE UNIT STEP FUNCTION

The **unit step function** (or **Heaviside function**) is defined by

$$H(t - a) = \begin{cases} 0 & t < a \\ 1 & t \geq a \end{cases} \tag{6.19}$$

and shown in Figure 6.2a. The unit step function can be used to construct other, more complicated step functions, for example $f(t) = H(t-a) - H(t-b)$ as illustrated in Figure 6.2b.

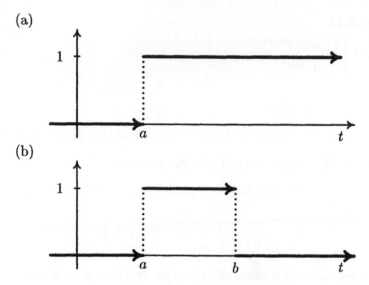

Figure 6.2: The unit step function. Shown are in (a) $H(t-a)$, and in (b) $H(t-a) - H(t-b)$, for $a < b$.

The Laplace transform of the unit step function is

$$L[H(t-a) : t \to s] = \int_0^\infty e^{-st} H(t-a)\, dt$$

$$= \int_a^\infty e^{-st}\, dt = \left[\frac{e^{-st}}{-s}\right]_a^\infty = \frac{e^{-as}}{s} \qquad (6.20)$$

valid for $s > 0$. The equation (6.20) provides the basis for the **second shifting property**. Let $\overline{f}(s)$ be the Laplace transform of $f(t)$ and suppose that $a > 0$ then

$$L[f(t-a)H(t-a) : t \to s] = e^{-as}\overline{f}(s) \qquad (6.21)$$

or, conversely,

$$L^{-1}\left[e^{-as}\overline{f}(s) : s \to t\right] = f(t-a)H(t-a). \qquad (6.22)$$

Proving the second shifting property is the task of Tutorial example T 6.12.

Example 6.11   Compute $L^{-1}\left[\dfrac{(1 - e^{-\pi s/2})}{1 + s^2} : s \to t\right]$.

Solution 6.11

$$L^{-1}\left[\frac{(1-e^{-\pi s/2})}{1+s^2}:s\to t\right]$$

$$= L^{-1}\left[\frac{1}{1+s^2}:s\to t\right] - L^{-1}\left[\frac{e^{-\pi s/2}}{1+s^2}:s\to t\right]$$

$$= \sin t - H(t-\pi/2)\sin(t-\pi/2) = \sin t + H(t-\pi/2)\cos t. \quad \square$$

Example 6.12   Solve the initial value problem

$$\ddot{y} + y = f(t), \qquad y(0) = 0, \quad \dot{y}(0) = 0,$$

with

$$f(t) = \begin{cases} 1 & 0 \le t < 1 \\ 0 & t \ge 1. \end{cases}$$

This problem arises, for example, when an undamped oscillator is disturbed from rest in its equilibrium state by a constant external driving force which is active for a finite period only.

Solution 6.12   The driving force $f(t)$ can be represented by the formula

$$f(t) = H(t) - H(t-1).$$

Let $\bar{y}(s) = L[y(t):t\to s]$, then by taking the Laplace transform of the differential equation it is seen that $\bar{y}(s)$ satisfies

$$\left[s^2\bar{y}(s) - s\dot{y}(0) - y(0)\right] + \bar{y}(s) = \frac{1}{s} - \frac{e^{-s}}{s}.$$

It is straightforward to verify that

$$\bar{y}(s) = \frac{1-e^{-s}}{s(s^2+1)} = \frac{1}{s} - \frac{s}{s^2+1} - \frac{e^{-s}}{s} + \frac{se^{-s}}{s^2+1}$$

where the details of the partial fraction decomposition to obtain the second expression have been omitted. Hence, after using the second shifting property on the final two terms,

$$y(t) = 1 - \cos t - H(t-1) + H(t-1)\cos(t-1)$$

$$= 1 - \cos t - [1 - \cos(t-1)]H(t-1).$$

The final expression can then be written as

$$y(t) = \begin{cases} 1 - \cos t & 0 \le t < 1 \\ \cos(t-1) - \cos t & t \ge 1. \end{cases}$$

It is worth noting that the solution is continuous at $t = 1$ even though the driving force $f(t)$ is not. $\qquad\qquad\square$

6.7 THE UNIT IMPULSE FUNCTION

In strict mathematical terms the **unit impulse function** (or **delta** or **Dirac delta** function), denoted by $\delta(t)$, is not an ordinary function but belongs to the class of so–called **generalized functions** or **distributions**. It is defined through the integral relation

$$\int_{-\infty}^{\infty} \delta(t-a)\, f(t)\, dt = f(a) \qquad (6.23)$$

for any integrable function $f(t)$.

Although the delta function is very special, its integral definition endows it with sufficiently good analytical properties that it can be used usefully in many mathematical applications almost as though it was a classical and not distributional function. These good analytical properties arise because the delta function may be thought of as a limit of a sequence of classical functions. Two examples illustrating this statement are now given.

Sequence 1: Let $\delta_\epsilon(t-a)$ be defined by the formula

$$\delta_\epsilon(t-a) = \begin{cases} \dfrac{1}{2\epsilon} & |t-a| < \epsilon \\[2mm] 0 & |t-a| \geq \epsilon. \end{cases} \qquad (6.24)$$

Therefore

$$\int_{-\infty}^{\infty} \delta_\epsilon(t-a)\, f(t)\, dt = \frac{1}{2\epsilon}\int_{a-\epsilon}^{a+\epsilon} f(t)\, dt = f(a+\theta\epsilon), \qquad |\theta| < 1$$

where the integrability of f has been used to invoke the **integral mean value theorem**. Thus

$$\lim_{\epsilon \to 0} \int_{-\infty}^{\infty} \delta_\epsilon(t-a)\, f(t)\, dt = f(a)$$

for any real integrable function f. In conclusion

$$\delta(t-a) = \lim_{\epsilon \to 0} \delta_\epsilon(t-a). \qquad (6.25)$$

Sequence 2: Another popular description of $\delta(t)$ (we have conveniently set $a = 0$ here) arises as the limit of the sequence

$$\delta_n(t) = \lim_{n \to \infty} \frac{n}{\sqrt{\pi}} e^{-n^2 t^2}. \qquad (6.26)$$

The function on the right–hand side of this equation is illustrated for different values of n in Figure 6.3.

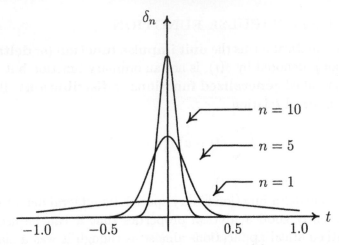

Figure 6.3: Graph of the function $(n/\sqrt{\pi})e^{-n^2t^2}$ for $n = 1, 5, 10$.

In the strict mathematical sense neither of these limits exists, but they provide useful conceptual ways towards an understanding of the delta function $\delta(t - a)$. By setting $f(t) = 1$ in (6.23), is evident that

$$\int_{-\infty}^{\infty} \delta(t - a)\, dt = 1. \qquad (6.27)$$

For this reason $\delta(t - a)$ is often called the *unit impulse function*. It can be described loosely as a function that is zero everywhere except at $t = a$. The basis for this perception is that in the definition (6.23) of $\delta(t - a)$, it is only the value of f at $t = a$ that contributes to the value of the integral.

Furthermore, it is clear that the unit step function $H(t - a)$ is related to the unit impulse function by

$$H(t - a) = \int_{-\infty}^{t} \delta(u - a)\, du \qquad (6.28)$$

or

$$\frac{d}{dt}\, H(t - a) = \delta(t - a). \qquad (6.29)$$

Once again, the correct interpretation of (6.28) and (6.29) is to be found in the theory of distributions. These formulas provide identities when both sides are multiplied with a function $f(t)$ and integrated with respect to t over the interval $(-\infty, \infty)$. Finally, the Laplace transform of $\delta(t - a)$ is

$$\begin{aligned} L\left[\delta(t - a) : t \to s\right] &= \int_{0}^{\infty} \delta(t - a)\, e^{-st}\, dt \\ &= \int_{-\infty}^{\infty} \delta(t - a)\, e^{-st}\, dt = e^{-as}, \quad (a > 0). \quad (6.30) \end{aligned}$$

Example 6.13   Solve the initial value problem

$$\ddot{y} + 2\dot{y} + y = \delta(t-1), \qquad y(0) = 2, \quad \dot{y}(0) = 3.$$

This equation represents a damped oscillator with an external driving force of infinite magnitude acting for an infinitely short time such that unit impulse is imparted to the system at time $t = 1$. Effectively, $\dot{y}$ jumps by one unit at $t = 1$.

Solution 6.13   Let $\overline{y}(s) = L[y(t) : t \to s]$ and take the Laplace transform of the differential equation. It follows that

$$\left[s^2\overline{y}(s) - 2s - 3\right] + 2\left[s\overline{y}(s) - 2\right] + \overline{y}(s) = e^{-s}$$

$$\longrightarrow \qquad \overline{y}(s) = \frac{2s + 7 + e^{-s}}{s^2 + 2s + 1} = \frac{2}{s+1} + \frac{5}{(s+1)^2} + \frac{e^{-s}}{(s+1)^2}$$

after taking partial fractions. Using both the first shifting property (6.5) and the second shifting property (6.22), the solution is seen to be

$$y(t) = 2e^{-t} + 5te^{-t} + (t-1)e^{-(t-1)}H(t-1). \qquad \square$$

A few further important Laplace transforms are listed in Table 6.2.

6.8   PERIODIC FUNCTIONS

Suppose that $f(t)$ is a real–valued function defined for $0 \le t < \infty$ with periodicity T. This means that for any non–negative value of t,

$$f(t) = f(t + kT), \qquad k = 1, 2, \ldots. \tag{6.31}$$

The Laplace transform of f becomes

$$L[f(t) : t \to s] = \overline{f}(s) = \int_0^\infty f(t)\,e^{-st}\,dt = \sum_{k=0}^\infty \int_{kT}^{(k+1)T} f(t)\,e^{-st}\,dt.$$

We use the substitution $t = u + kT$ in these integrals and exploit the periodicity of f. We then see from the previous equation that

$$\overline{f}(s) = \sum_{k=0}^\infty \int_0^T f(u + kT)\,e^{-s(u+kT)}\,du = \sum_{k=0}^\infty e^{-kTs} \int_0^T f(u)\,e^{-su}\,du.$$

The infinite series can be summed up in closed form. Hence if $f(t)$ is a real–valued function with periodicity T then

$$\overline{f}(s) = \frac{1}{1 - e^{-sT}} \int_0^T f(t)\,e^{-st}\,dt. \tag{6.32}$$

	$f(t) = L^{-1}\left[\overline{f}(s) : s \to t\right]$	$\overline{f}(s) = L\left[f(t) : t \to s\right]$	
12.	$H(t-a)$	$\dfrac{e^{-as}}{s}$	$\mathrm{Re}(s) > 0$
13.	$H(t-a)f(t-a)$	$e^{-as}\overline{f}(s)$	
14.	$e^{at}f(t)$	$\overline{f}(s-a)$	
15.	$f(at)$	$\dfrac{1}{a}\overline{f}\left(\dfrac{s}{a}\right)$	$a > 0$
16.	$\displaystyle\int_0^t f(t-w)g(w)\,dw$	$\overline{f}(s)\overline{g}(s)$	
17.	$\delta(t-a)$	e^{-as}	
18.	$f^{(n)}(t)$	$s^n\overline{f}(s) - \displaystyle\sum_{j=0}^{n-1} s^{n-j-1}f^{(j)}(0)$	
19.	$(-t)^n f(t)$	$\overline{f}^{(n)}(s)$	

Table 6.2: More Laplace transforms.

Example 6.14 Demonstrate the validity of the formula (6.32) for the Laplace transform of a real–valued function of periodicity T by using it to calculate the Laplace transform of $\sin at$.

Solution 6.14 The Laplace transform of $\sin at$ has already been shown to be $a/(a^2+s^2)$. We begin by recognizing that $\sin at$ has periodicity $T = 2\pi/a$, and consequently $\sin aT = 0$, $\cos aT = 1$. Therefore

$$\int_0^T e^{-st}\sin at\,dt = \left[\frac{-e^{-st}}{s}\sin at\right]_0^T + \frac{a}{s}\int_0^T e^{-st}\cos at\,dt$$

$$= \frac{a}{s}\left[\frac{-e^{-st}}{s}\cos at\right]_0^T - \frac{a^2}{s^2}\int_0^T e^{-st}\sin at\,dt$$

$$= \frac{a}{s^2}\left[1 - e^{-sT}\right] - \frac{a^2}{s^2}\int_0^T e^{-st}\sin at\,dt\,.$$

Thus,

$$\int_0^T e^{-st} \sin at\, dt = \frac{a(1 - e^{-sT})}{a^2 + s^2},$$

and application of (6.32) now gives the Laplace transform of $\sin at$.   □

Common periodic waveforms

The *saw tooth* illustrated in Figure 6.4a and the *square wave* illustrated in Figure 6.4b are two common periodic waveforms worthy of our attention.

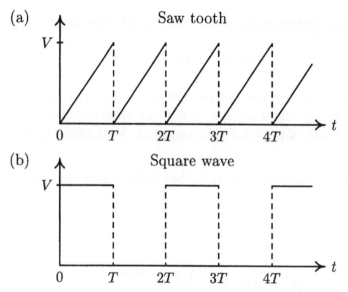

Figure 6.4: Part (a) illustrates a saw tooth waveform of amplitude V and period T while part (b) illustrates a square wave pulse of amplitude V and period $2T$.

Saw tooth. The Laplace transform of the saw tooth waveform of amplitude V and period T is calculated from (6.32) by recognizing first that

$$\int_0^T f(t)\, e^{-st}\, dt = \frac{V}{T} \int_0^T t e^{-st}\, dt = \frac{V}{T} \frac{1}{s^2} \left(1 - e^{-sT} - sT e^{-sT}\right).$$

Thus the Laplace transform of the saw tooth wave is seen from (6.32) to be

$$\overline{f}(s) = \frac{V}{T} \frac{1}{s^2} \left(1 - \frac{sT e^{-sT}}{1 - e^{-sT}}\right). \tag{6.33}$$

Square wave. The Laplace transform of the square wave of amplitude V and period $2T$ is calculated from (6.32) by the formula

$$\overline{f}(s) = \frac{V}{1 - e^{-2sT}} \int_0^T e^{-st}\, dt = \frac{V}{s}\frac{1}{1 + e^{-sT}}. \qquad (6.34)$$

Other periodic pulses. We briefly mention a further four commonly used periodic functions (amplitude V) that are formed by repeating the pulses

Meander function

$$f(t) = \begin{cases} V & t \in (0,T) \\ -V & t \in (T, 2T) \end{cases}$$

Triangular wave

$$f(t) = \begin{cases} tV/T & t \in (0,T) \\ (2T - t)V/T & t \in (T, 2T) \end{cases}$$

Full wave rectification

$$f(t) = \{ \ V\sin(\pi t/T) \quad t \in (0,T)$$

Half wave rectification

$$f(t) = \begin{cases} V\sin(\pi t/T) & t \in (0,T) \\ 0 & t \in (T, 2T). \end{cases}$$

Example 6.15 Determine the current in a series circuit containing an inductor of L Henry and a resistor of R Ohm when the circuit is driven by a saw tooth voltage of period T and amplitude V. Assume that no current is flowing initially.

Solution 6.15 The current $I(t)$ in the circuit is governed by the differential equation

$$L\dot{I} + RI = U_{\text{saw}}(t), \qquad I(0) = 0.$$

The Laplace transform $\overline{I}(s) = L[I(t) : t \to s]$ therefore satisfies the algebraic equation

$$(Ls + R)\overline{I} = \frac{V}{T}\frac{1}{s^2}\left(1 - \frac{sTe^{-sT}}{1 - e^{-sT}}\right).$$

Define $a = R/L$ then $\overline{I}$ is given by

$$\begin{aligned}
\overline{I} &= \frac{V}{RT}\frac{a}{s^2(s+a)}\left(1 - \frac{sTe^{-sT}}{1 - e^{-sT}}\right) \\
&= \frac{V}{RT}\frac{1}{s^2}\left(1 - \frac{sTe^{-sT}}{1 - e^{-sT}}\right) - \frac{V}{RT}\frac{1}{s(s+a)}\left(1 - \frac{sTe^{-sT}}{1 - e^{-sT}}\right).
\end{aligned}$$

In order to interpret meaningfully the Laplace transform $\overline{I}$, we recognize that the first component in the last expression is a multiple of $L[U_{\text{saw}}(t) : t \to s]$, see (6.33). To be explicit,

$$\begin{aligned}
\overline{I} - \frac{\overline{U}_{\text{saw}}}{R} &= -\frac{VL}{R^2 T}\frac{a}{s(s+a)} + \frac{V}{R}\frac{1}{(s+a)}\frac{e^{-sT}}{1 - e^{-sT}} \\
&= -\frac{VL}{R^2 T}\frac{a}{s(s+a)} + \frac{V}{R}\sum_{k=1}^{\infty}\frac{e^{-skT}}{(s+a)}.
\end{aligned}$$

Using the inverse Laplace transforms

$$L^{-1}\left[\frac{a}{s(s+a)} : s \to t\right] = 1 - e^{-at},$$

$$L^{-1}\left[\frac{e^{-skT}}{(s+a)} : s \to t\right] = e^{-a(t-kT)}\, H(t-kT),$$

in which H is the Heaviside function, it is now seen that

$$I(t) = \frac{U_{\text{saw}}(t)}{R} - \frac{VL}{R^2 T}\left(1 - e^{-at}\right) + \frac{V}{R}\sum_{k=1}^{\infty} e^{-a(t-kT)}\, H(t-kT). \quad \square$$

6.9   TUTORIAL EXAMPLES 6

T 6.1   Find the Laplace transform of $f(t) = e^{(-t-1/2)}$.

T 6.2   Find the Laplace transform of $f(t) = \cos(at + b)$.

T 6.3   Calculate the Laplace transform of $f(t) = t^p$ (for $p > -1$). Express the solution in terms of the **Gamma function** $\Gamma(x)$ defined by

$$\Gamma(x) = \int_0^\infty u^{x-1} e^{-u}\, du, \quad x > 0.$$

Consider four special cases: p is an integer; $p = -1/2$; $p = 1/2$; $p = 5/2$.
[*Hint*: make use of the formulas $\Gamma(x+1) = x\Gamma(x)$; $\Gamma(1/2) = \pi^{1/2}$].

T 6.4   Find the inverse Laplace transform of $\overline{f}(s) = (s-2)/(s^2 - 2)$.

T 6.5   Find the inverse Laplace transform of $\overline{f}(s) = 3/(s^2 + 4s + 9)$.

T 6.6   Solve the convolution integral equation

$$f(t) = 1 + \int_0^t f(u)\cos(t-u)\, du.$$

T 6.7   Solve the convolution integral equation

$$\sin t - t = \int_0^t (t-u)^2 f(u)\, du.$$

T 6.8   Use Laplace transforms to solve the initial value problem

$$\ddot{y}(t) - 5\dot{y}(t) + 6y(t) = 0, \quad y(0) = 2, \quad \dot{y}(0) = 1.$$

T 6.9   Use Laplace transforms to solve the initial value problem

$$\ddot{y}(t) - y(t) = te^{2t}, \qquad y(0) = 0, \quad \dot{y}(0) = 1.$$

T 6.10   Write $f(t)$ as a step function and calculate its Laplace transform

$$f(t) = \begin{cases} 1 & t < 1 \\ 3 & 1 \le t < 7 \\ 5 & t \ge 7. \end{cases}$$

T 6.11   Use Laplace transforms to solve the initial value problem

$$4\ddot{y}(t) + 4\dot{y}(t) + 5y(t) = g(t), \qquad y(0) = \dot{y}(0) = 0$$

$$g(t) = \begin{cases} 4 & 0 \le t < \pi \\ 0 & t \ge \pi \end{cases}$$

and sketch the solution $y(t)$.

T 6.12   Let $\overline{f}(s)$ be the Laplace transform of $f(t)$. By using the definition of the Laplace transform, prove the second shifting property (6.21), namely

$$L[f(t-a)H(t-a) : t \to s] = e^{-as}\overline{f}(s)$$

where H is the Heaviside function and $a > 0$.

T 6.13   Solve the initial value problem

$$\ddot{y}(t) + 2\dot{y}(t) + 2y(t) = \delta(t - \pi), \qquad y(0) = \dot{y}(0) = 0$$

by using Laplace transforms and sketch the solution $y(t)$.

T 6.14   Consider an electrical circuit where an inductor of inductance L and a capacitor of capacitance C are connected in series. Before the circuit is activated, there is a charge of Q_0 on the capacitor. At $t = 0$ the circuit is activated. Formulate and solve the initial value problem for the charge $Q(t)$ on the capacitor by means of Laplace transform. Interpret the solution. Return to the more general serial LCR circuit by adding a resistor of resistance R to the previous circuit. Obtain the solution of the corresponding initial value problem for $R^2 < 4L/C$ and, in particular, compare the solution in the limit $R^2 \ll 4L/C$ with the earlier result. Can you recognize the physical mechanism behind the terms in the solution?

T 6.15   Determine the current in a series circuit containing an inductor of L Henry and a resistor of R Ohm when the circuit is driven by a square wave voltage of period $2T$ and amplitude V. Assume that no current is flowing initially.

7

Higher Order Initial Value Problems

7.1   GENERAL REMARKS

A linear differential equation for $y(x)$ of order n is defined by

$$p_n(x)y^{(n)}(x) + p_{n-1}(x)y^{(n-1)}(x) + \ldots + p_1(x)y'(x) + p_0(x)y(x) = g(x). \quad (7.1)$$

Once again this equation is called *inhomogeneous* if $g(x) \neq 0$ and *homogeneous* if $g(x) \equiv 0$. The general solution of (7.1) contains n integration constants and to obtain a unique solution for an initial value problem, n independent initial conditions must therefore be specified. These are typically

$$y(x_0) = y_0, \quad y'(x_0) = y_0', \quad \ldots \quad , \quad y^{(n-1)}(x_0) = y_0^{(n-1)} \quad (7.2)$$

(note that the primes on the constants y_0', etc., are just notational conventions and do not signify derivatives). Let $y_1(x), y_2(x), \ldots, y_n(x)$ be n solutions of the homogeneous equation, then the linear combination

$$y(x) = c_1 y_1(x) + c_2 y_2(x) + \ldots + c_n y_n(x) \quad (7.3)$$

where $c_1, c_2, \ldots, c_n$, are constants, is also a solution. Now, define the *Wronskian* or *Wronskian determinant*

$$W(y_1, y_2, \ldots, y_n) = \begin{vmatrix} y_1 & y_2 & \cdots & y_n \\ y_1' & y_2' & \cdots & y_n' \\ \vdots & \vdots & \ddots & \vdots \\ y_1^{(n-1)} & y_2^{(n-1)} & \cdots & y_n^{(n-1)} \end{vmatrix}. \quad (7.4)$$

When each row of the Wronskian determinant is differentiated with respect to x, the row beneath it is generated *except* for the last row. Bearing in mind that a determinant with two identical rows is zero, it therefore follows that

$$\frac{dW}{dx} = \begin{vmatrix} y_1 & y_2 & \cdots & y_n \\ y_1' & y_2' & \cdots & y_n' \\ \vdots & \vdots & \ddots & \vdots \\ y_1^{(n-2)} & y_2^{(n-2)} & \cdots & y_n^{(n-2)} \\ y_1^{(n)} & y_2^{(n)} & \cdots & y_n^{(n)} \end{vmatrix}. \tag{7.5}$$

Values for $y_k^{(n)}(x)$ on the right–hand side of (7.5) are replaced from the original differential equation. All derivatives of order $(n-2)$ and below in $y_k^{(n)}(x)$ generate multiples of the first $(n-1)$ rows of W and therefore make no contribution to the value of W'. The only component of $y_k^{(n)}(x)$ to contribute to W' comes from the coefficient $-p_{n-1}(x)/p_n(x)$ of $y_k^{(n-1)}(x)$. Therefore, the Wronskian $W(y_1, y_2, \ldots, y_n)$ satisfies the differential equation

$$p_n(x)W' + p_{n-1}(x)W = 0 \quad \longrightarrow \quad W(x) = W(x_0)e^{-\int_{x_0}^x \frac{p_{n-1}(s)}{p_n(s)} ds}. \tag{7.6}$$

Since the exponential function is always positive, it follows from (7.6) that if W is zero at any point, then it must be identically zero since $W(x_0) \equiv 0$. Thus, the Wronskian $W(y_1, \ldots, y_n)$ is either identically zero for all $x \geq x_0$ or it is never zero. If $W \not\equiv 0$, the set of solutions is called a **fundamental set of solutions**. The functions $y_1, \ldots, y_n$ are then **linearly independent** and every solution of the homogeneous version of equation (7.1) can be expressed in the form (7.3). On the other hand, if $W \equiv 0$ then the set of solutions is incomplete in the sense that $y_1, \ldots, y_n$ are **linearly dependent**.

The solution to the inhomogeneous equation (7.1) is given by

$$y(x) = c_1 y_1(x) + c_2 y_2(x) + \ldots + c_n y_n(x) + y_p(x) \tag{7.7}$$

where $y_p(x)$ is a particular integral of the inhomogeneous equation (7.1).

Example 7.1   Verify that the functions $1, x, \cos x, \sin x$ are solutions of the differential equation $y^{(4)} + y'' = 0$ and determine their Wronskian.

Solution 7.1

$y:$	1	x	$\cos x$	$\sin x$
$y':$	0	1	$-\sin x$	$\cos x$
$y'':$	0	0	$-\cos x$	$-\sin x$
$y''':$	0	0	$\sin x$	$-\cos x$
$y^{(4)}:$	0	0	$\cos x$	$\sin x$

Simple addition shows that the four functions indeed fulfil the differential equation. The Wronskian is

$$W(y_1, y_2, y_3, y_4) = \begin{vmatrix} 1 & x & \cos x & \sin x \\ 0 & 1 & -\sin x & \cos x \\ 0 & 0 & -\cos x & -\sin x \\ 0 & 0 & \sin x & -\cos x \end{vmatrix}$$

$$= \begin{vmatrix} 1 & -\sin x & \cos x \\ 0 & -\cos x & -\sin x \\ 0 & \sin x & -\cos x \end{vmatrix} = \begin{vmatrix} -\cos x & -\sin x \\ \sin x & -\cos x \end{vmatrix}$$

$$= \cos^2 x - (-\sin^2 x) = 1.$$

Therefore the functions $1, x, \cos x, \sin x$ form a fundamental set of solutions and the general solution of $y^{(4)} + y'' = 0$ is

$$y(x) = c_1 + c_2 x + c_3 \cos x + c_4 \sin x.$$ □

7.2   EQUATIONS WITH CONSTANT COEFFICIENTS

A linear, homogeneous differential equation of order n with constant coefficients has general form

$$a_n y^{(n)} + a_{n-1} y^{(n-1)} + a_{n-2} y^{(n-2)} + \cdots + a_1 y' + a_0 y = 0. \qquad (7.8)$$

In the following, the constants $a_1, \ldots, a_n$, are assumed to be complex–valued and $a_n \neq 0$. Equation (7.8) has solutions of form $e^{\lambda x}$ provided λ satisfies the *auxiliary equation*

$$p(\lambda) = a_n \lambda^n + a_{n-1} \lambda^{n-1} + \ldots + a_1 \lambda + a_0 = 0 \qquad (7.9)$$

where $p(\lambda)$ is a polynomial of degree n. Suppose that the n roots of $p(\lambda)$ are $\lambda_1, \ldots, \lambda_n$ then the auxiliary equation can be factorized as

$$p(\lambda) = a_n (\lambda - \lambda_1)(\lambda - \lambda_2) \ldots (\lambda - \lambda_n) = a_n \prod_{k=1}^{n} (\lambda - \lambda_k). \qquad (7.10)$$

Two different cases are possible: roots can have multiplicity $s = 1$ or $s > 1$.

Case 1: When all roots λ_k are distinct (i.e., they all have multiplicity 1), the general solution of (7.8) is

$$y = c_1 e^{\lambda_1 x} + c_2 e^{\lambda_2 x} + \ldots + c_n e^{\lambda_n x}. \qquad (7.11)$$

In the particular instance in which a_k $(0 \leq k \leq n)$ are real–valued, complex conjugate pairs $\lambda = \alpha \pm i\beta$ can arise. Terms in the general solution (7.11) may now be regrouped into combinations of $e^{\alpha x} \cos \beta x$ and $e^{\alpha x} \sin \beta x$ (compare this with the analogous procedure for second order equations in Section 3.3). **Case 2:** If roots are repeated and if, say, λ_k is a repeated root of multiplicity $s > 1$, then the s functions

$$ e^{\lambda_k x}, \quad x e^{\lambda_k x}, \quad x^2 e^{\lambda_k x}, \quad \ldots, \quad x^{s-1} e^{\lambda_k x} \tag{7.12} $$

are all solutions of (7.8) whereas each root λ_j of multiplicity 1 contributes a term $e^{\lambda_j x}$ to the solution. Likewise if the coefficients of (7.8) are all real–valued, the terms in (7.12) may be regrouped into $e^{\alpha x} \cos \beta x$ and $e^{\alpha x} \sin \beta x$ terms.

Example 7.2   Find the general solution of the fourth order differential equation $y^{(4)} - y = 0$.

Solution 7.2   The auxiliary equation is $\lambda^4 - 1 = (\lambda^2 - 1)(\lambda^2 + 1) = 0$, which has the four distinct roots $1, -1, i, -i$. Therefore

$$ y(x) = c_1 e^x + c_2 e^{-x} + c_3 e^{ix} + c_4 e^{-ix} $$

or, equivalently,

$$ y = c_1 e^x + c_2 e^{-x} + c_5 \sin x + c_6 \cos x $$

is the general solution.   □

Example 7.3   Find the general solution of $y^{(4)} + 2y'' + y = 0$.

Solution 7.3   The auxiliary equation is $\lambda^4 + 2\lambda^2 + 1 = (\lambda^2 + 1)^2 = 0$, which has roots $i, i, -i, -i$. Both i and $-i$ occur as repeated roots with multiplicity 2. This leads to the general solution

$$ y(x) = c_1 e^{ix} + c_2 e^{-ix} + c_3 x e^{ix} + c_4 x e^{-ix}, $$

more commonly expressed in the form

$$ y(x) = c_5 \cos x + c_6 \sin x + x(c_7 \cos x + c_8 \sin x). $$ □

Of course, the roots of polynomials of degree n $(n \geq 5)$ cannot be found by a general formula. Therefore in practice this strategy is extremely limited. The method succeeds only when the auxiliary equation has roots that can be found by inspection or guess–work.

7.3   THE INHOMOGENEOUS EQUATION

Solutions of the inhomogeneous equation of order n $(a_n \neq 0)$

$$a_n y^{(n)} + a_{n-1} y^{(n-1)} + a_{n-2} y^{(n-2)} + \cdots + a_1 y' + a_0 y = g(x) \qquad (7.13)$$

can now be obtained by the techniques of Chapter 4 in an analogous way.

Undetermined coefficients

To make a correct selection for the particular integral, we generalize the procedure outlined in Section 4.2 to differential equations of order n. Let $\hat{H}_n$ be the linear differential operator defined by

$$\hat{H}_n(y) = a_n y^{(n)} + a_{n-1} y^{(n-1)} + \cdots + a_1 y' + a_0 y = \sum_{k=0}^{n} a_k \, y^{(k)}$$

and let $p_n(\lambda) = 0$ be the auxiliary equation associated with $\hat{H}_n$ where

$$p_n(\lambda) = a_0 + a_1 \lambda + \ldots + a_n \lambda^n = \sum_{k=0}^{n} a_k \lambda^k .$$

For any complex constant α and suitably differentiable function $h(x)$, **Leibniz' theorem** gives

$$\hat{H}_n \Big[h(x) e^{\alpha x} \Big] = \sum_{k=0}^{n} a_k \left[\sum_{m=0}^{k} \binom{k}{m} \alpha^{k-m} h^{(m)} \right] e^{\alpha x}$$

where $\binom{k}{m}$ is the *binomial coefficient*. The order of summation in this expression is now reversed to obtain

$$\hat{H}_n \Big[h(x) e^{\alpha x} \Big] = \sum_{m=0}^{n} h^{(m)} \left(\sum_{k=m}^{n} \binom{k}{m} a_k \alpha^{k-m} \right) e^{\alpha x}$$

$$= \sum_{m=0}^{n} h^{(m)} \frac{1}{m!} \frac{d^m p_n(\lambda)}{d\lambda^m} \bigg|_{\lambda=\alpha} e^{\alpha x} .$$

However, if α is a root of $p_n(\lambda) = 0$ of multiplicity s then

$$p_n(\alpha) = \frac{dp_n(\lambda)}{d\lambda} \bigg|_{\lambda=\alpha} = \ldots = \frac{d^{s-1} p_n(\lambda)}{d\lambda^{s-1}} \bigg|_{\lambda=\alpha} = 0$$

and in these circumstances,

$$\hat{H}_n \Big[h(x) e^{\alpha x} \Big] = \sum_{m=s}^{n} h^{(m)} \frac{1}{m!} \frac{d^m p_n(\lambda)}{d\lambda^m} \bigg|_{\lambda=\alpha} e^{\alpha x} . \qquad (7.14)$$

For example, in order to treat an inhomogeneous equation whose right–hand side has a term of form $g(x)e^{\alpha x}$ where $g(x)$ is a polynomial in x of degree r and α is a zero of the auxiliary equation with multiplicity s, expression (7.14) reveals that the s–th derivative of h must have order r. Thus h must be a polynomial of order $r + s$ and may be represented without loss of generality in the form $h(x) = x^s q(x)$ where $q(x)$ is an arbitrary polynomial of degree r. Therefore, Table 4.1 applies again with s chosen according to the developments of the current paragraph.

Variation of constants

Application of the method of variation of constants to an inhomogeneous differential equation of higher order is explored in considerable detail in Tutorial example T 7.7.

Exact equations

The criterion for an exact equation, which we presented for second order equations in Section 4.7, can be generalized to higher order equations. The differential equation

$$\sum_{k=0}^{n} a_k(x)\, y^{(k)}(x) = f(x) \qquad (7.15)$$

is *exact* if and only if the coefficients $a_k(x)$ satisfy the condition

$$\sum_{k=0}^{n} (-1)^k\, a_k^{(k)}(x) = 0\,. \qquad (7.16)$$

Integration of (7.15) then yields

$$\sum_{k=0}^{n-1} \left[\sum_{m=0}^{n-k-1} (-1)^m\, a_{m+k+1}^{(m)}(x) \right] y^{(k)}(x) = \int^{x} f(s)\, ds\,. \qquad (7.17)$$

which constitutes a differential equation or order $n - 1$.

Reduction of order

Suppose that $w(x)$ is any solution of the homogeneous equation corresponding to the inhomogeneous equation

$$\sum_{k=0}^{n} a_k(x)\, y^{(k)}(x) = f(x)\,, \qquad (7.18)$$

then the change of dependent variable from y to v defined by $y(x) = w(x)v(x)$ leads to a linear differential equation for v of order $n - 1$, that is, of order one less than that for y. The equation for v is

$$\sum_{k=0}^{n} a_k(x) (wv)^{(k)}(x) = \sum_{k=0}^{n} a_k(x) \sum_{r=0}^{k} \binom{k}{r} v^{(r)}(x) w^{(k-r)}(x) = f(x). \quad (7.19)$$

When the order of summation is changed in this double sum, the equation for v becomes

$$
\begin{aligned}
f(x) &= \sum_{r=0}^{n} v^{(r)}(x) \sum_{k=r}^{n} a_k(x) \binom{k}{r} w^{(k-r)}(x) \\
&= v(x) \sum_{k=0}^{n} a_k(x) w^{(k)}(x) + \sum_{r=1}^{n} v^{(r)}(x) \sum_{k=r}^{n} a_k(x) \binom{k}{r} w^{(k-r)}(x) \\
&= \sum_{r=1}^{n} v^{(r)}(x) \sum_{k=r}^{n} a_k(x) \binom{k}{r} w^{(k-r)}(x)
\end{aligned}
$$

since $w(x)$ is given to satisfy the homogeneous form of equation (7.18). Now let $z = v'$ then z satisfies

$$
\begin{aligned}
f(x) &= \sum_{r=1}^{n} z^{(r-1)}(x) \sum_{k=r}^{n} a_k(x) \binom{k}{r} w^{(k-r)}(x) \\
&= \sum_{r=0}^{n-1} z^{(r)}(x) \sum_{k=r}^{n-1} a_{k+1}(x) \binom{k+1}{r+1} w^{(k-r)}(x) \quad (7.20)
\end{aligned}
$$

which is a linear differential equation of order $n - 1$.

The most interesting and powerful application of this idea arises when $n = 2$ — the specialization covered in Section 4.5 — since the method now yields a first order equation which may fall into one of the many categories of equations already discussed.

Remarks 7.1   In preparation for the next example we note that a differential equation

$$\sum_{k=0}^{n} a_k(x) y^{(k)}(x) = 0$$

has at least one solution of the form $y(x) = e^{\lambda x}$ if

$$\sum_{k=0}^{n} a_k(x) \lambda^k = 0.$$

Example 7.4   Find the general solution of the differential equation

$$x^2 y''' - x^2 y'' - 2y' + 2y = 0.$$

Solution 7.4   By using the result of Remarks 7.1 we see that the coefficient functions of this equation sum to zero. Therefore $e^{\lambda x}$, for $\lambda = 1$, is a solution of the homogeneous equation. Now write $y = e^x v$ where v is the new dependent variable then

$$x^2 \left(v''' + 3v'' + 3v' + v \right) - x^2 \left(v'' + 2v' + v \right) - 2\left(v' + v - v \right) = 0$$

which simplifies to

$$x^2 v''' + 2x^2 v'' + (x^2 - 2)v' = 0 \qquad \longrightarrow \qquad x^2 z'' + 2x^2 z' + (x^2 - 2)z = 0$$

where $z = v'$. This equation is not exact by criterion (7.16) but upon multiplication with the integrating factor e^x we see that

$$e^x \left[x^2 z'' + 2x^2 z' + (x^2 - 2)z \right] = \frac{d}{dx} \left[x^2 e^x z' + (x^2 - 2x)e^x z \right] = 0.$$

Now the equation is exact and can integrated to

$$x^2 e^x z' + (x^2 - 2x)e^x z = A.$$

This last equation has standard form

$$z' + \left(1 - \frac{2}{x} \right) z = A \frac{e^{-x}}{x^2}.$$

The differential equation for z is linear with integrating factor e^x / x^2. Thus

$$\frac{d}{dx} \left(\frac{e^x}{x^2} z \right) = \frac{A}{x^4}$$

with solution

$$\frac{e^x}{x^2} z = -\frac{A}{3x^3} + B \qquad \longrightarrow \qquad z = -A \frac{e^{-x}}{3x} + Bx^2 e^{-x}.$$

A further integration now yields

$$v(x) = \int z(x)\, dx = -\frac{A}{3} \int \frac{e^{-x}}{x}\, dx - B(x^2 + 2x + 2)e^{-x} + C.$$

In conclusion, the general solution of the original equation is

$$y(x) = a\, e^x \int^x \frac{e^{-s}}{s}\, ds + b(x^2 + 2x + 2) + c\, e^x$$

where a, b and c are arbitrary constants. Note that the existence of the solution at $x = 0$ requires $a = 0$ to remove the divergent integral.   □

Laplace transform

Once again, the Laplace transform provides an attractive method for the solution of initial value problems of higher order. The key formula that opens the way to successful application is

$$L\left[f^{(n)}(t) : t \to s\right] = s^n \overline{f}(s) - \sum_{k=0}^{n-1} s^{n-k-1} f^{(k)}(0). \qquad (7.21)$$

which gives the Laplace transform of the n–th derivative of a function $f(t)$. Expression (7.21) is a straightforward generalization of (6.13) and (6.14) for the first and second derivative respectively and can be proved analogously.

7.4   TUTORIAL EXAMPLES 7

T 7.1   Obtain three functions y_1, y_2 and y_3 that are solutions of the differential equation $xy''' - y'' = 0$ and determine their Wronskian.

T 7.2   Obtain three functions y_1, y_2 and y_3 that are solutions of the differential equation $x^3 y''' + x^2 y'' - 2xy' + 2y = 0$ and determine their Wronskian.

T 7.3   Find the general solution of $y^{(4)} + y = 0$.

T 7.4   Find the general solution of $y^{(4)} - 8y' = 0$.

T 7.5   Find the general solution of $y^{(4)} - 2y'' + y = 0$.

T 7.6   Solve the initial value problem

$$y''' - 3y'' + 2y' = x + e^x, \quad y(0) = 1, \ y'(0) = -1/4, \ y''(0) = -3/2.$$

by using the *method of undetermined coefficients*.
[*Hint*: in order to find a suitable expression for the particular integral use Table 4.1 and interpret s as discussed in Section 7.3.]

T 7.7   Find the general solution of the differential equation

$$y''' + y' = \tan x, \qquad 0 < x < \pi/2$$

by using the method of *variation of constants*. Follow the procedure:

  1. Find a fundamental set of solutions (y_1, y_2, y_3) of the homogeneous equation.

2. Set

$$y_p(x) = u_1(x)y_1(x) + u_2(x)y_2(x) + u_3(x)y_3(x)$$

to obtain a particular integral of the inhomogeneous equation. Substituting this into the differential equation would give only one equation for the three unknown functions u_1, u_2, u_3.

3. It appears that two more conditions are needed to determine u_1, u_2 and u_3. These can be conveniently chosen as

$$y_1 u_1' + y_2 u_2' + y_3 u_3' = 0,$$
$$y_1' u_1' + y_2' u_2' + y_3' u_3' = 0$$

in order to avoid higher order derivatives of u_1, u_2, u_3 in y_p'' and y_p'''.

4. Show then that

$$u_m' = (\tan x)\, W_m(x)/W(x), \qquad m = 1, 2, 3$$

where $W(x) = W(y_1, y_2, y_3)(x)$ is the Wronskian and W_m is the determinant obtained from W by replacing the m-th column by the column $(0, 0, 1)^T$.

5. Integrate u_m' to obtain $u_m(x)$.

Generalize the procedure to a differential equation of order n with an arbitrary inhomogeneous term $g(x)$ on the right–hand side.

T 7.8   Use Laplace transforms to solve the initial value problem

$$y^{(4)}(t) - y(t) = \delta(t - 1), \qquad y(0) = \dot{y}(0) = \ddot{y}(0) = y^{(3)}(0) = 0.$$

8

Systems of First Order Linear Equations

8.1 INTRODUCTION

Assume that instead of a single function $y(x)$, it is now necessary to find N unknown functions, denoted by $y_1(x)$, $y_2(x)$, ..., $y_N(x)$. A **system of first order linear differential equations** has general form

$$
\begin{aligned}
y_1'(x) &= a_{11}(x)y_1(x) + a_{12}(x)y_2(x) + \ldots + a_{1N}(x)y_N(x) + b_1(x) \\
y_2'(x) &= a_{21}(x)y_1(x) + a_{22}(x)y_2(x) + \ldots + a_{2N}(x)y_N(x) + b_2(x) \\
&\ \ \vdots \qquad\qquad\qquad\qquad\qquad\ \ \vdots \\
y_n'(x) &= a_{n1}(x)y_1(x) + a_{n2}(x)y_2(x) + \ldots + a_{nN}(x)y_N(x) + b_n(x) \\
&\ \ \vdots \qquad\qquad\qquad\qquad\qquad\ \ \vdots \\
y_N'(x) &= a_{N1}(x)y_1(x) + a_{N2}(x)y_2(x) + \ldots + a_{NN}(x)y_N(x) + b_N(x) \, .
\end{aligned}
\tag{8.1}
$$

These equations may be rewritten in the more compact notation

$$
y_k'(x) = \sum_{n=1}^{N} a_{kn}(x)y_n(x) + b_k(x), \quad k = 1,2,\ldots,N \, .
\tag{8.2}
$$

A well defined initial value problem for $x \geq x_0$ arises when the system of equations (8.1) or (8.2) is supplemented by N initial conditions

$$
y_k(x_0) = y_{k0}, \quad k = 1,2,\ldots,N \, .
\tag{8.3}
$$

Translation into **matrix notation** yields[1]

$$
\mathbf{Y}'(x) = \mathbf{A}(x)\mathbf{Y}(x) + \mathbf{B}(x), \qquad \mathbf{Y}(x_0) = \mathbf{Y}_0
\tag{8.4}
$$

---

[1]The matrices and column vectors that appear in this chapter are generally denoted by capital letters in bold face.

where $\mathbf{Y}(x)$, $\mathbf{B}(x)$ and $\mathbf{Y}_0$ are the column vectors (length N)

$$\mathbf{Y}(x) = \begin{bmatrix} y_1(x) \\ y_2(x) \\ \vdots \\ y_N(x) \end{bmatrix}, \quad \mathbf{B}(x) = \begin{bmatrix} b_1(x) \\ b_2(x) \\ \vdots \\ b_N(x) \end{bmatrix}, \quad \mathbf{Y}_0 = \begin{bmatrix} y_{10} \\ y_{20} \\ \vdots \\ y_{N0} \end{bmatrix} \qquad (8.5)$$

and $\mathbf{A}(x)$ is the $N \times N$ matrix function given by

$$\mathbf{A}(x) = \begin{bmatrix} a_{11}(x) & a_{12}(x) & \cdots & a_{1N}(x) \\ a_{21}(x) & a_{22}(x) & \cdots & a_{2N}(x) \\ \vdots & \vdots & \ddots & \vdots \\ a_{N1}(x) & a_{N2}(x) & \cdots & a_{NN}(x) \end{bmatrix}. \qquad (8.6)$$

Two techniques to deal with such systems are studied. However, it is clear that any generalized analytical approach necessarily needs to make specific assumptions concerning the functions $a_{kn}(x)$ and $b_k(x)$. Suppose for the time being that the system of first order equations has *constant coefficients*. In this case

$$a_{kn}(x) = a_{kn}, \quad b_k(x) = b_k, \quad k, n = 1, 2, \ldots, N \qquad (8.7)$$

or alternatively, $\mathbf{A}$ is a constant matrix and $\mathbf{B}$ is a constant vector.

8.2　ELIMINATION

Consider first the case $N = 2$, i.e., two equations for two unknown functions $y_1(x)$ and $y_2(x)$. The system (8.1) with the simplifications (8.7) reduces to

$$\begin{align} y_1' &= a_{11}y_1 + a_{12}y_2 + b_1 \qquad &(8.8) \\ y_2' &= a_{21}y_1 + a_{22}y_2 + b_2 \qquad &(8.9) \end{align}$$

and two initial conditions $y_1(x_0) = y_{10}$ and $y_2(x_0) = y_{20}$ will in general be specified. Differentiation of (8.8) and substitution for y_2' from (8.9) gives

$$y_1'' = a_{11}y_1' + a_{12}y_2' = a_{11}y_1' + a_{12}(a_{21}y_1 + a_{22}y_2 + b_2).$$

Next, eliminate y_2 from the second term in the bracket by solving (8.8) explicitly for y_2 in terms of y_1 and y_1'. The result is

$$y_1'' = a_{11}y_1' + a_{12}a_{21}y_1 + a_{12}b_2 + a_{22}(y_1' - a_{11}y_1 - b_1)$$

or, tidied up,

$$y_1'' - (a_{11} + a_{22})y_1' + (a_{11}a_{22} - a_{12}a_{21})y_1 = a_{12}b_2 - a_{22}b_1. \qquad (8.10)$$

Equation (8.10) is now a standard second order differential equation for y_1. Let λ_1 and λ_2 be the roots of the auxiliary equation

$$\lambda^2 - (a_{11} + a_{22})\lambda + (a_{11}a_{22} - a_{12}a_{21}) = 0 \qquad (8.11)$$

then

$$y_1(t) = y_{1c}(t) + y_{1p}(t) \qquad (8.12)$$

where the complementary function y_{1c} is given by

$$y_{1c}(t) = \begin{cases} D_1 e^{\lambda_1 x} + D_2 e^{\lambda_2 x} & \lambda_1 \neq \lambda_2 \\ (D_1 + D_2 x)\, e^{\lambda_1 x} & \lambda_1 = \lambda_2. \end{cases} \qquad (8.13)$$

The particular integral $y_{1p}(t)$ can now be obtained by the method of undetermined coefficients. Once $y_1(x)$ is known, $y_2(x)$ is normally computed from (8.8). This is a straightforward operation, in that knowledge of y_1 yields y_2 without further integration. Indeed, if $a_{12} \neq 0$, then

$$y_2(x) = \frac{1}{a_{12}}\left(y_1' - a_{11}y_1 - b_1\right). \qquad (8.14)$$

It is easy to see that this method, while still relatively easy for the case $N = 2$, becomes more tedious and intractable as N increases.

Remarks 8.1

1. It has just been demonstrated that two linear differential equations of first order involving two unknown functions normally lead to one linear differential equation of second order for one unknown function. This equivalence extends generally in that N linear differential equations of first order for N unknown functions are equivalent to one linear differential equation of N-th order for one unknown function (from which the remaining $N-1$ unknown functions can then be calculated). Although justified here for linear equations only, the result is true generally.

2. The converse result is also true, namely, given a single differential equation of order N, there is a family of N independent variables (the choice of which is not unique) such that the original equation can be re–expressed as a system of N first order equations. Such manipulations of the original equation are central to the numerical solution of ordinary differential equations. Indeed, the vast majority of schemes for the numerical solution of differential equations take it for granted that the

equations are formulated as a first order system. As an example of one popular strategy, take the third order equation

$$y''' + p_1(x)y'' + p_2(x)y' + p_3(x)y = q(x)$$

with initial values $y(x_0) = y_0$, $y'(x_0) = y_0'$, $y''(x_0) = y_0''$. The equation can be re–expressed as a system of three first order equations using the three new functions defined by

$$w_1(x) = y(x), \quad w_2(x) = y'(x), \quad w_3(x) = y''(x).$$

Self–evidently, w_1, w_2 and w_3 satisfy

$$\begin{aligned} w_1' &= w_2 \\ w_2' &= w_3 \\ w_3' &= -p_1 w_3 - p_2 w_2 - p_3 w_1 + q \end{aligned}$$

with initial conditions $w_1(x_0) = y_0$, $w_2(x_0) = y_0'$, $w_3(x_0) = y_0''$.

3. It was noticed in connection with (8.14) that y_2 can normally be determined without further integration, although (8.14) holds only if $a_{12} \neq 0$. If $a_{12} = 0$, however, (8.8) and (8.9) can be solved sequentially without recourse to the elaborate elimination procedure. One solves (8.8) as an independent first order equation for y_1 and then solves (8.9) for y_2. This is exactly the procedure that was employed earlier in Section 3.4, see (3.16), to obtain the solution of the inhomogeneous equation of second order.

4. The integration constants D_1 and D_2 in (8.13) can be determined directly from (8.12), i.e., before even calculating y_2 in (8.14). In order to do so we require $y_1(x_0)$ and $y_1'(x_0)$. While $y_1(x_0)$ is explicitly given, $y_1'(x_0)$ can be calculated by substituting the given initial values $y_1(x_0)$ and $y_2(x_0)$ into (8.8). $\triangle$

Example 8.1   Solve the system of first order differential equations

$$2y_1' = -y_1 + 4y_2, \quad 2y_2' = -4y_1 - y_2$$

with the initial conditions $y_1(0) = -2$, $y_2(0) = 2$.

Solution 8.1   Differentiating the first of the two equations gives

$$2y_1'' = -y_1' + 4y_2' = -y_1' + 2(-4y_1 - y_2)$$

where the second equation has been used to eliminate y_2'. From the first equation, we have $4y_2 = 2y_1' + y_1$. Substituting this expression into the previous equation and rearranging terms yields

$$4y_1'' + 4y_1' + 17y_1 = 0 .$$

The auxiliary equation is $\lambda^2 + \lambda + 17/4 = 0$ with roots $\lambda_{1,2} = -(1/2) \pm 2i$. Therefore, the general solution for y_1 is

$$y_1(x) = e^{-x/2} \left(A \cos 2x + B \sin 2x \right) .$$

Using $4y_2 = 2y_1' + y_1$ provides

$$y_2(x) = e^{-x/2} \left(B \cos 2x - A \sin 2x \right) .$$

The initial conditions are satisfied by $A = -2$ and $B = 2$, and the particular solution of this initial value problem is given by

$$y_1(x) = 2e^{-x/2} \left(\sin 2x - \cos 2x \right) ,$$
$$y_2(x) = 2e^{-x/2} \left(\sin 2x + \cos 2x \right) . \qquad \square$$

Example 8.2   The trajectory $x = x(t)$, $y = y(t)$ of a golf ball of mass m struck with initial speed v_0 and rising initially at angle θ_0 satisfies the differential equations

$$m\ddot{x} = -R_x , \qquad\qquad m\ddot{y} = -mg - R_y$$

with initial conditions

$$x(0) = y(0) = 0 , \qquad \dot{x}(0) = v_0 \cos \theta_0 , \quad \dot{y}(0) = v_0 \sin \theta_0 .$$

In these equations, g (assumed constant) is the gravitational acceleration, $x(t)$ and $y(t)$ are the horizontal range and vertical height of the ball at time t, and R_x, R_y are respectively the horizontal and vertical components of air resistance.

(a) Write down the given initial value problem as a fourth order system using the dependent variables $y_1(t) = x(t)$, $y_2(t) = y(t)$, $y_3(t) = \dot{x}(t)$ and $y_4(t) = \dot{y}(t)$.

(b) What is the trajectory of a golf ball assuming that air resistance is proportional to velocity, that is, $R_x = mk\dot{x}$, $R_y = mk\dot{y}$.

Solution 8.2   (a) If

$$y_1(t) = x(t), \quad y_2(t) = y(t), \quad y_3(t) = \dot{x}(t), \quad y_4(t) = \dot{y}(t)$$

then, from these definitions it follows immediately that $\dot{y}_1 = y_3$ and $\dot{y}_2 = y_4$. Furthermore,

$$\dot{y}_3 = \ddot{x} = -\frac{R_x}{m}, \qquad \dot{y}_4 = \ddot{y} = -g - \frac{R_y}{m}.$$

Therefore the initial value problem may now be restated as the first order system

$$
\begin{aligned}
\dot{y}_1 &= y_3 & y_1(0) &= 0 \\
\dot{y}_2 &= y_4 & y_2(0) &= 0 \\
\dot{y}_3 &= -R_x/m & y_3(0) &= v_0 \cos \theta_0 \\
\dot{y}_4 &= -g - R_y/m & y_4(0) &= v_0 \sin \theta_0.
\end{aligned}
$$

(b) The model for air resistance gives $R_x = mky_3$ and $R_y = mky_4$. In this case the initial value problem becomes

$$
\begin{aligned}
\dot{y}_1 &= y_3 & y_1(0) &= 0 \\
\dot{y}_2 &= y_4 & y_2(0) &= 0 \\
\dot{y}_3 &= -ky_3 & y_3(0) &= v_0 \cos \theta_0 \\
\dot{y}_4 &= -g - ky_4 & y_4(0) &= v_0 \sin \theta_0.
\end{aligned}
$$

The third and fourth equations can be integrated to give

$$y_3(t) = v_0 \cos \theta_0 \, e^{-kt},$$
$$y_4(t) = -\frac{g}{k} + \left(\frac{g}{k} + v_0 \sin \theta_0 \right) e^{-kt}.$$

Once y_3 and y_4 are known then y_1 and y_2 can be determined by two further integrations. Specifically,

$$y_1(t) = \frac{v_0 \cos \theta_0}{k} \left(1 - e^{-kt} \right),$$
$$y_2(t) = -\frac{gt}{k} + \left(\frac{g}{k^2} + \frac{v_0 \sin \theta_0}{k} \right) \left(1 - e^{-kt} \right).$$

The one–dimensional motion analysed in Section 2.3 appears as a specialization of this problem when $\theta_0 = \pi/2$. Indeed, the expression for $y_4(t) = \dot{y}(t)$ in this example is identical to (2.20) when $\theta_0 = \pi/2$ for $t_0 = 0$.   □

8.3   MATRIX METHOD

Return now to the general problem for N unknown functions and consider the special case in which the coefficient matrix $\mathbf{A}$ is constant. The matrix formulation of the initial value problem is

$$\frac{d\mathbf{Y}(x)}{dx} = \mathbf{A}\mathbf{Y}(x) + \mathbf{B}(x), \quad \mathbf{Y}(x_0) = \mathbf{Y}_0. \tag{8.15}$$

This equation has general solution of the form

$$\mathbf{Y}(x) = \mathbf{Y}_c(x) + \mathbf{Y}_p(x) \tag{8.16}$$

where $\mathbf{Y}_c(x)$ — the complementary function — is the solution of the homogeneous equation

$$\mathbf{Y}'(x) = \mathbf{A}\mathbf{Y}(x) \tag{8.17}$$

whereas $\mathbf{Y}_p(x)$ is a particular integral. Let $\mathbf{Y}_c(x) = \mathbf{Y}_* e^{\lambda x}$, where $\mathbf{Y}_*$ is a constant vector of length N, and λ be a constant parameter, then

$$\mathbf{A}\mathbf{Y}_c - \mathbf{Y}'_c = (\mathbf{A} - \lambda \mathbf{I})\mathbf{Y}_* e^{\lambda x}$$

where $\mathbf{I}$ is the identity matrix of order N. Therefore $\mathbf{Y}_* e^{\lambda x}$ is a solution of the homogeneous equation (8.17) if and only if $\mathbf{Y}_*$ is a non–zero solution of the equation

$$(\mathbf{A} - \lambda \mathbf{I})\mathbf{Y}_* = \mathbf{0}. \tag{8.18}$$

This is possible if and only if $(\mathbf{A} - \lambda \mathbf{I})$ is singular, that is, λ satisfies

$$\det(\mathbf{A} - \lambda I) = 0. \tag{8.19}$$

The last equation is a polynomial[2] of degree N with N roots $\lambda_1, \lambda_2, \ldots, \lambda_N$ commonly called the **eigenvalues** of $\mathbf{A}$. Each eigenvalue λ_k is accompanied by a non–zero $\mathbf{Y}_{*k}$ called an **eigenvector**. The construction of $\mathbf{Y}_c$ has therefore been reduced to a **matrix eigenvalue problem**.

If the eigenvalues are all distinct then the corresponding N eigenvectors are linearly independent and the complementary function is

$$\mathbf{Y}_c(x) = \sum_{k=1}^{N} \mathbf{Y}_{*k} e^{\lambda_k x} \tag{8.20}$$

where $\mathbf{Y}_{*k}$ is an eigenvector corresponding to λ_k. Otherwise, $\mathbf{A}$ has repeated eigenvalues and the previous expression for the complementary function is

---

[2]Of course, (8.19) is the auxiliary equation of the equivalent N–th order linear differential equation that would have been obtained by eliminating all unknown functions except one, see (8.11) for the case $N = 2$.

deficient. The situation is analogous to that experienced when the auxiliary equation has repeated roots — a case discussed in detail in Sections 3.3 and 7.2. For example, if λ_1 is a double eigenvalue of $\mathbf{A}$ then the contribution to the particular integral takes the form

$$(x\mathbf{Y}_1 + \mathbf{Y}_2)e^{\lambda_1 x}$$

where

$$(\mathbf{A} - \lambda_1\mathbf{I})\mathbf{Y}_1 = \mathbf{0}, \qquad \mathbf{Y}_1 = (\mathbf{A} - \lambda_1\mathbf{I})\mathbf{Y}_2.$$

Thus $\mathbf{Y}_2$ is the general solution of $(\mathbf{A} - \lambda_1\mathbf{I})^2\mathbf{Y}_2 = \mathbf{0}$ while $\mathbf{Y}_1$ is computed from $\mathbf{Y}_2$ by the formula $\mathbf{Y}_1 = (\mathbf{A} - \lambda_1\mathbf{I})\mathbf{Y}_2$. In this instance, $\mathbf{Y}_2$ has two degrees of freedom. Suppose that $\mathbf{A}$ has r distinct eigenvalues $\lambda_1, \ldots, \lambda_r$ with multiplicities $k_1, \ldots, k_r$ (in this instance $k_1 = 2$) then the **Cayley–Hamilton theorem** states that

$$\prod_{j=1}^{r}(\mathbf{A} - \lambda_j\mathbf{I})^{k_j} = \mathbf{0}.$$

If $\mathbf{Z}$ is a vector of length N whose entries $z_1, \ldots, z_N$, are all arbitrary constants then $\mathbf{Y}_2$ may be formally represented by

$$\mathbf{Y}_2 = \left[\prod_{j=2}^{r}(\mathbf{A} - \lambda_j\mathbf{I})^{k_j}\right]\mathbf{Z}.$$

The following simple example illustrates the method.

Example 8.3   Express the equations

$$y_1' = -5y_1 + 4y_2, \qquad y_2' = -9y_1 + 7y_2$$

in matrix form and determine the eigenvalues of the system matrix. Compute the general solution of this system of equations.

Solution 8.3   The equations have form

$$\frac{d\mathbf{Y}}{dx} = \mathbf{AY}, \qquad \mathbf{Y} = \begin{bmatrix} y_1 \\ y_2 \end{bmatrix}, \qquad \mathbf{A} = \begin{bmatrix} -5 & 4 \\ -9 & 7 \end{bmatrix}.$$

The eigenvalues of $\mathbf{A}$ are solutions of

$$\begin{vmatrix} -5 - \lambda & 4 \\ -9 & 7 - \lambda \end{vmatrix} = \lambda^2 - 2\lambda + 1 = (\lambda - 1)^2 = 0.$$

Thus $\lambda_{1,2} = 1$ is a double root. By calculation

$$(\mathbf{A} - \lambda\mathbf{I}) = \begin{bmatrix} -6 & 4 \\ -9 & 6 \end{bmatrix} \qquad (\mathbf{A} - \lambda\mathbf{I})^2 = \begin{bmatrix} 0 & 0 \\ 0 & 0 \end{bmatrix}.$$

Therefore

$$\mathbf{Y}_2 = \begin{bmatrix} c_1 \\ c_2 \end{bmatrix}, \qquad \mathbf{Y}_1 = \begin{bmatrix} -6 & 4 \\ -9 & 6 \end{bmatrix}\begin{bmatrix} c_1 \\ c_2 \end{bmatrix} = \begin{bmatrix} -6c_1 + 4c_2 \\ -9c_1 + 6c_2 \end{bmatrix}$$

and the general solution is

$$\mathbf{Y} = (x\mathbf{Y}_1 + \mathbf{Y}_2)e^x = \begin{bmatrix} (-6c_1 + 4c_2)xe^x + c_1 e^x \\ (-9c_1 + 6c_2)xe^x + c_2 e^x \end{bmatrix},$$

or in a more familiar format,

$$y_1(x) = 2(2c_2 - 3c_1)xe^x + c_1 e^x, \quad y_2(x) = 3(2c_2 - 3c_1)xe^x + c_2 e^x. \quad \square$$

The calculation of particular integrals is again complicated and not described in any detail at this point except to remark that if $\mathbf{A}$ is non–singular and $\mathbf{B}(x)$ is a constant vector then

$$\mathbf{Y}_p(x) = -\mathbf{A}^{-1}\mathbf{B} \tag{8.21}$$

is a particular integral of (8.15) where $\mathbf{A}^{-1}$ is the inverse matrix of $\mathbf{A}$.

In general, there are exactly N integration constants embedded in the arbitrariness of eigenvectors. These constants are determined for an initial value problem by solving a system of N linear equations formed from the initial condition $\mathbf{Y}(x_0) = \mathbf{Y}_0$.

8.4 MATRIX EXPONENT METHOD

The notion of a **matrix exponent** provides a more elegant solution of the initial value problem

$$\frac{d\mathbf{Y}(x)}{dx} = \mathbf{A}(x)\mathbf{Y}(x) + \mathbf{B}(x), \qquad \mathbf{Y}(x_0) = \mathbf{Y}_0, \tag{8.22}$$

when $\mathbf{A}(x)$ and $\mathbf{B}(x)$ are prescribed matrix functions of x and not necessarily constant. The method relies on the linearity of (8.22) and integrates the system of equations using a **matrix integrating factor**.

Let $\mathbf{A}$ be a $N \times N$ matrix, then the exponent of $\mathbf{A}$, denoted by $\exp \mathbf{A}$,[3] is the $N \times N$ matrix defined by

$$\exp \mathbf{A} = \mathbf{I} + \mathbf{A} + \frac{\mathbf{A}^2}{2} + \ldots + \frac{\mathbf{A}^k}{k!} + \ldots = \sum_{k=0}^{\infty} \frac{\mathbf{A}^k}{k!} \qquad (8.23)$$

where $\mathbf{A}^2 = \mathbf{A}\mathbf{A}$, etc. The convergence of the power series for e^x for all values of x guarantees that the series for $\exp \mathbf{A}$ converges for all matrices $\mathbf{A}$. Some immediate consequences of the definition (8.23) are now listed.

1. The exponent of the zero $N \times N$ matrix $\mathbf{0}$ is the $N \times N$ identity matrix $\mathbf{I}$, i.e. $\exp \mathbf{0} = \mathbf{I}$.

2. If $\mathbf{A}$ is a constant matrix then

$$\frac{d \exp(\mathbf{A}x)}{dx} = \frac{d}{dx} \left(\sum_{k=0}^{\infty} \frac{\mathbf{A}^k x^k}{k!} \right) = \sum_{k=1}^{\infty} \frac{\mathbf{A}^k x^{k-1}}{(k-1)!} = \mathbf{A} \exp(\mathbf{A}x). \quad (8.24)$$

3. $\exp \mathbf{A}$ commutes with any power of $\mathbf{A}$ — the proof is by inspection.

4. If $\mathbf{B}$ commutes with $\mathbf{A}$, i.e., $\mathbf{A}\mathbf{B} = \mathbf{B}\mathbf{A}$, then $\mathbf{B}$ commutes with $\exp \mathbf{A}$.

5. If $\mathbf{A}$ and $\mathbf{B}$ are $N \times N$ matrices then

$$\exp \mathbf{A} \exp \mathbf{B} = \sum_{k=0}^{\infty} \frac{\mathbf{A}^k}{k!} \sum_{j=0}^{\infty} \frac{\mathbf{B}^j}{j!} = \sum_{n=0}^{\infty} \frac{1}{n!} \left[\sum_{k=0}^{n} \binom{n}{k} \mathbf{A}^k \mathbf{B}^{n-k} \right]. \quad (8.25)$$

The sum contained in the square brackets is $(\mathbf{A} + \mathbf{B})^n$ only when $\mathbf{A}$ and $\mathbf{B}$ are **commuting matrices**. Therefore, the relation

$$\exp \mathbf{A} \exp \mathbf{B} = \exp(\mathbf{A} + \mathbf{B}) = \exp \mathbf{B} \exp \mathbf{A} \qquad (8.26)$$

holds only for commuting matrices $\mathbf{A}$ and $\mathbf{B}$.

6. Since $\mathbf{A}$ and $-\mathbf{A}$ commute then

$$\exp \mathbf{A} \exp(-\mathbf{A}) = \exp \mathbf{0} = \mathbf{I} = \exp(-\mathbf{A}) \exp \mathbf{A}. \qquad (8.27)$$

Thus $\exp \mathbf{A}$ has inverse $\exp(-\mathbf{A})$ for any matrix $\mathbf{A}$.

---

[3]We use $\exp \mathbf{A}$ rather than our standard notation $e^{\mathbf{A}}$ as a reminder that the definition is purely formal.

The concept of the exponent of a matrix can now be employed to solve (8.22). Let matrix $\mathbf{M}(x)$ be the solution of the matrix equation

$$\frac{d\mathbf{M}(x)}{dx} = -\mathbf{M}(x)\mathbf{A}(x), \qquad \mathbf{M}(x_0) = \mathbf{I}. \qquad (8.28)$$

If $\mathbf{Y}(x)$ satisfies (8.22) then

$$\frac{d(\mathbf{MY})}{dx} = \mathbf{M}\frac{d\mathbf{Y}}{dx} + \frac{d\mathbf{M}}{dx}\mathbf{Y} = \mathbf{M}(\mathbf{AY} + \mathbf{B}) - (\mathbf{MA})\mathbf{Y} = \mathbf{MB}. \qquad (8.29)$$

By formal integration of this equation (the right–hand side contains known functions) it follows that

$$
\begin{aligned}
\mathbf{M}(x)\mathbf{Y}(x) &= \int_{x_0}^{x} \mathbf{M}(u)\mathbf{B}(u)\, du + \mathbf{M}(x_0)\mathbf{Y}(x_0) \\
&= \int_{x_0}^{x} \mathbf{M}(u)\mathbf{B}(u)\, du + \mathbf{Y}_0 \qquad (8.30)
\end{aligned}
$$

where the integral of a matrix is simply the matrix of integrated elements. Provided $\mathbf{M}(x)$ is non–singular for all x, then problem (8.22) has solution

$$\mathbf{Y}(x) = \mathbf{M}^{-1}(x) \int_{x_0}^{x} \mathbf{M}(u)\mathbf{B}(u)\, du + \mathbf{M}^{-1}(x)\mathbf{Y}_0. \qquad (8.31)$$

It can be shown that $\mathbf{M}(x)$ exists for all $\mathbf{A}$ and is invertible. The argument is based on the iterative construction

$$\mathbf{M}_0(x) = \mathbf{I}, \qquad \mathbf{M}_{k+1}(x) = \int_{x_0}^{x} \mathbf{M}_k(s)\mathbf{A}(s)\, ds, \qquad k = 0, 1, \dots$$

which in turn leads to the definition

$$\mathbf{M}(x) = \sum_{k=0}^{\infty} (-1)^{k+1}\, \mathbf{M}_k(x).$$

It may be demonstrated that $\mathbf{M}(x)$ satisfies (8.28) although the details of the argument are not pursued here.

For the special case in which $\mathbf{A}$ is a constant matrix, it is evident from the properties of $\exp \mathbf{A}$ that

$$\mathbf{M}(x) = \exp\left[-\mathbf{A}(x - x_0)\right] \qquad (8.32)$$

and in this case, (8.22) has solution

$$\mathbf{Y}(x) = \int_{x_0}^{x} \left[\exp \mathbf{A}(x - u)\right] \mathbf{B}(u)\, du + \exp\left[\mathbf{A}(x - x_0)\right] \mathbf{Y}_0. \qquad (8.33)$$

Example 8.4   Use the matrix exponent method to determine the general solution of the system of differential equations given in Example 8.3.

Solution 8.4   The system given in Example 8.3 has the matrix form

$$\frac{d\mathbf{Y}}{dx} = \mathbf{A}\mathbf{Y}, \qquad \mathbf{Y} = \begin{bmatrix} y_1 \\ y_2 \end{bmatrix}, \qquad \mathbf{A} = \begin{bmatrix} -5 & 4 \\ -9 & 7 \end{bmatrix}.$$

The general solution is therefore $\mathbf{Y} = \exp(\mathbf{A}x)\,\mathbf{C}$ where $\mathbf{C}$ is a vector of two arbitrary constants. It has already been shown that $(\mathbf{A} - \mathbf{I})^2 = \mathbf{0}$. Clearly

$$\mathbf{Y} = \exp\left[(\mathbf{A} - \mathbf{I})x + \mathbf{I}x\right] \mathbf{C} = \exp\left[(\mathbf{A} - \mathbf{I})x\right] \exp(\mathbf{I}x)\,\mathbf{C}$$

where it has been recognized that $\mathbf{I}$ and $(\mathbf{A} - \mathbf{I})$ are commuting matrices. However,

$$\exp(\mathbf{I}x) = e^x\,\mathbf{I}, \qquad \exp\left[(\mathbf{A} - \mathbf{I})x\right] = \sum_{k=0}^{\infty} \frac{x^k(\mathbf{A} - \mathbf{I})^k}{k!} = \mathbf{I} + x(\mathbf{A} - \mathbf{I})$$

because $(\mathbf{A} - \mathbf{I})^k = \mathbf{0}$ for $k \geq 2$. Therefore, the general solution is

$$\mathbf{Y} = e^x\,\mathbf{C} + x\,e^x\,(\mathbf{A} - \mathbf{I})\mathbf{C}$$

with component form

$$y_1(x) = 2(2c_2 - 3c_1)x\,e^x + c_1\,e^x, \qquad y_2(x) = 3(2c_2 - 3c_1)x\,e^x + c_2\,e^x.$$

This is exactly the same solution as that determined previously. Furthermore, the choice $\mathbf{C} = \mathbf{Y}_0$ gives a solution that would fulfil the initial condition $\mathbf{Y}(0) = \mathbf{Y}_0$ (with $x_0 = 0$).   □

8.5   TUTORIAL EXAMPLES 8

T 8.1   Solve the initial value problem

$$y_1'(x) = y_1(x) - 2y_2(x), \qquad y_1(0) = -1,$$
$$y_2'(x) = 3y_1(x) - 4y_2(x), \qquad y_2(0) = 2.$$

T 8.2   A primitive **predator–prey model** for foxes and hares assumes that the fox population $f(t)$ increases at a rate proportional to the size of the hare population $h(t)$ and is reduced by a mortality rate of μ_f. Similarly, the hare population declines at a rate proportional to the size of the fox population but is sustained by a large birth rate μ_h ($> \mu_f$). For some period the fox and hare populations have been static at F and H respectively. A

recent outbreak of *leveret myxomatosis* rapidly reduced the hare population to fraction k of its previous level thus creating an ecological imbalance.
Construct differential equations for f and h and hence deduce a second order
differential equation for h. Use this model to predict the future development
of the fox and hare populations.

T 8.3 Take the equations of Example 8.2 as a description of the initial
value problem for the trajectory of a golf ball in a resisting medium and
assume that

$$R_x = mf(v)\,\frac{\dot{x}}{v}, \qquad R_y = mf(v)\,\frac{\dot{y}}{v}$$

where $f(v)$ is the magnitude of the air resistance at speed $v = \sqrt{\dot{x}^2 + \dot{y}^2}$.
Note that v which denotes velocity elsewhere, refers to speed here and thus
$v \geq 0$.

1. If the golf ball is at (x, y) at time t and is travelling at speed $v(t)$ and
   at inclination $\theta(t)$ $(-\pi/2 < \theta < \pi/2$ for all times $t)$ to the horizontal,
   show that the differential equations from Example 8.2 can be replaced
   by the system

$$
\begin{aligned}
\dot{x} &= v\cos\theta   & x(0) &= 0 \\
\dot{y} &= v\sin\theta   & y(0) &= 0 \\
\dot{v} &= -f(v) - g\sin\theta & v(0) &= v_0 \\
\dot{\theta} &= -\frac{g\cos\theta}{v} & \theta(0) &= \theta_0 .
\end{aligned}
$$

   It should be noted that this system of first order equations is non–
   linear.

2. Assuming that air resistance is proportional to the speed itself, that
   is, $f(v) = kgv$ calculate the inclination of the golf ball at any time.

3. If air resistance is constant, show that the golf ball speed at trajectory
   angle θ satisfies

$$v = \frac{v^*(\cos\theta)^{k-1}}{(1 - \sin\theta)^k}$$

   where v^* is the speed of the ball at the top of its flight–path.

9

Boundary Value Problems

9.1 MOTIVATION

We consider the vibration of an **elastic string** stretched to tension T between two fixed points which are distance l apart as schematically shown in Figure 9.1.

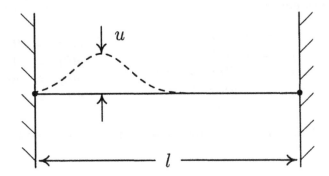

Figure 9.1: Schematic graph of the vibrations of an elastic string of length l with u being the vertical displacement.

Let $u(x, t)$ be the vertical displacement of the string at point x and at time t, then the **equation of motion** of the string is

$$\frac{\partial^2 u}{\partial x^2} = \frac{1}{c^2} \frac{\partial^2 u}{\partial t^2} \qquad (9.1)$$

where $c^2 = T/\rho$ and ρ is the string mass per unit length. The equation (9.1) is a **partial differential equation**, or, to be more specific, a **wave equation** in time and one space dimension. Partial differential equations of this type appear often in many applications of mathematics but their general treatment goes beyond the scope of this book.

The full description of the problem requires the specification of **initial conditions**

$$u(x, t_0) = r(x), \qquad \frac{\partial u(x, t)}{\partial t}\bigg|_{t=t_0} = s(x) \qquad (9.2)$$

describing the shape of the string and its velocity at time $t = t_0$ through two functions $r(x)$ and $s(x)$ respectively; and **boundary conditions**

$$u(0, t) = 0, \qquad u(l, t) = 0 \qquad (9.3)$$

describing the fixed nature of the boundary points for all times t.

Many partial differential equations such as (9.1) can be solved by assuming that the solution is a product of functions, each one depending on only one variable, a technique commonly called **separation of variables** for partial differential equations. The functional form

$$u(x, t) = F(x)\, G(t) \qquad (9.4)$$

satisfies (9.1) if $F(x)$ and $G(t)$ fulfil

$$\frac{d^2 F}{dx^2} + \alpha F = 0, \qquad \frac{d^2 G}{dt^2} + c^2 \alpha G = 0 \qquad (9.5)$$

for some arbitrary constant α. In view of (9.4), the boundary conditions (9.3) for $u(x, t)$ are fulfilled by

$$F(0) = 0, \qquad F(l) = 0. \qquad (9.6)$$

The first equation of (9.5) and (9.6) define a **boundary value problem** for $F(x)$, namely

$$F''(x) + \alpha F(x) = 0, \qquad F(0) = 0, \quad F(l) = 0. \qquad (9.7)$$

The general solution of the differential equation in (9.7) for $\alpha > 0$ is

$$F(x) = A \sin \sqrt{\alpha}\, x + B \cos \sqrt{\alpha}\, x. \qquad (9.8)$$

Applying the boundary conditions $F(0) = 0$ and $F(l) = 0$ yields

$$0 = B, \quad 0 = A \sin \sqrt{\alpha}\, l + B \cos \sqrt{\alpha}\, l \quad \longrightarrow \quad B = 0, \quad A \sin \sqrt{\alpha}\, l = 0.$$

Since B is already zero then the choice $A = 0$ produces only the trivial solution $F(x) \equiv 0$. A non–trivial solution ($A \neq 0$) emerges if and only if

$$\sin \sqrt{\alpha}\, l = 0 \qquad (9.9)$$

which requires that

$$\sqrt{\alpha_n}\, l = n\pi \qquad \longrightarrow \qquad \alpha_n = \frac{n^2\pi^2}{l^2}, \qquad n = 1, 2, 3, \ldots . \qquad (9.10)$$

Thus $\alpha_1, \alpha_2, \ldots$ form a family of values of α such that (9.7) has a non–trivial solution. Each value of α is called an **eigenvalue**, and the corresponding expression for $F(x)$ is called an **eigenfunction**. The eigenfunctions of (9.7) are therefore

$$F_n(x) = \sin\frac{n\pi x}{l}, \qquad n = 1, 2, 3, \ldots \qquad (9.11)$$

and the linear combination

$$F(x) = \sum_{n=1}^{\infty} a_n F_n(x) = \sum_{n=1}^{\infty} a_n \sin\frac{n\pi x}{l} \qquad (9.12)$$

is a general solution of the boundary value problem (9.7) for arbitrary constants a_n.

Remarks 9.1

1. The physical interpretation of $F_n(x) = \sin(n\pi x/l)$ is that, for each n, the function $F_n(x)$ denotes a fundamental vibration (a so–called normal mode) of the string.

2. Problem (9.7) has the trivial solution for all values of α other than those given by (9.10).

3. The solution (9.12) still contains arbitrary constants a_n. This presents no difficulty since F forms only part of the solution for $u(x,t)$; equation (9.1) with (9.2) and (9.3) has a unique solution.

4. In writing down the solution (9.8) it was assumed explicitly that the arbitrary constant $\alpha > 0$. If $\alpha < 0$, expression (9.8) for F is replaced by

$$F(x) = Ae^{\sqrt{-\alpha}x} + Be^{-\sqrt{-\alpha}x}.$$

The boundary conditions require $A+B = 0$ and $Ae^{\sqrt{-\alpha}l}+Be^{-\sqrt{-\alpha}l} = 0$ and thus

$$\begin{bmatrix} 1 & 1 \\ e^{\sqrt{-\alpha}l} & e^{-\sqrt{-\alpha}l} \end{bmatrix}\begin{bmatrix} A \\ B \end{bmatrix} = \begin{bmatrix} 0 \\ 0 \end{bmatrix}.$$

The determinant of this linear system has value $-2\sinh\left(\sqrt{-\alpha}l\right)$, which is never zero. Thus $A = B = 0$ is the only possible solution and therefore $F(x) \equiv 0$. $\triangle$

9.2   BOUNDARY VALUE PROBLEMS OF SECOND ORDER

A general **boundary value problem of second order** consists of the differential equation

$$y''(x) + p(x)y'(x) + q(x)y(x) = h(x), \quad a < x < b, \tag{9.13}$$

and two boundary conditions, one specified at $x = a$ and the other at $x = b$. These boundary conditions may be of different type. They can be of the **first kind**

$$y(a) = \eta_1, \quad y(b) = \eta_2, \tag{9.14}$$

of the **second kind**

$$y'(a) = \eta_1, \quad y'(b) = \eta_2, \tag{9.15}$$

of the **third kind**

$$\alpha_1 y(a) + \alpha_2 y'(a) = \eta_1, \quad \beta_1 y(b) + \beta_2 y'(b) = \eta_2, \tag{9.16}$$

or they can be of **periodic type**

$$y(a) = y(b), \quad y'(a) = y'(b) \tag{9.17}$$

where α_1, α_2, β_1, β_2, η_1, η_2 are constants. It is not difficult to see that the first and the second kind are both special cases of the third kind.

Example 9.1   Find the positive values of λ for which the boundary value problem $y'' - 2y' + (1 + \lambda)y = 0$, $y(0) = 0$, $y(1) = 0$ has non–trivial solutions.

Solution 9.1   With $y = e^{\alpha x}$ we have $\alpha^2 - 2\alpha + (1 + \lambda) = 0$, which yields

$$\alpha_{1,2} = 1 \pm \sqrt{1 - (1 + \lambda)} = 1 \pm i\sqrt{\lambda}.$$

Thus the complementary function is

$$y(x) = Ae^{(1+i\sqrt{\lambda})x} + Be^{(1-i\sqrt{\lambda})x}$$

or

$$y(x) = e^x(C \cos \sqrt{\lambda}\, x + D \sin \sqrt{\lambda}\, x).$$

The boundary conditions require

$$\begin{aligned} y(0) &= 0 & \longrightarrow & \quad C = 0, \\ y(1) &= 0 & \longrightarrow & \quad e^1\left(C \cos \sqrt{\lambda} + D \sin \sqrt{\lambda}\right) = 0, \end{aligned}$$

which can only provide a non–trivial solution when

$$\sqrt{\lambda_n} = n\pi, \qquad n = 1, 2, 3, \ldots.$$

The solutions of the boundary value problem are therefore the functions

$$y_n(x) = k_n\, e^x \sin n\pi x, \qquad n = 1, 2, 3, \ldots$$

where k_n are arbitrary constants.   □

Example 9.2   Find the values of the parameter λ for which the boundary value problem $y'' + \lambda y = 0$, $y'(0) = 0$, $y(\pi) = 0$, has non–trivial solutions.

Solution 9.2   We begin by establishing if the boundary value problem can be solved for $\lambda = 0$. For $\lambda = 0$, the differential equation simplifies to $y'' = 0$ which has the solution $y = ax + b$. Substitution of the two boundary conditions shows that $a = b = 0$. This reduces the solution to the trivial solution. Therefore we have

$$y(x) = A \sin \sqrt{\lambda}\, x + B \cos \sqrt{\lambda}\, x, \qquad \lambda \neq 0.$$

The boundary conditions require

$$\begin{aligned} y'(0) &= 0 &&\longrightarrow && A\sqrt{\lambda} = 0, \\ y(\pi) &= 0 &&\longrightarrow && A \sin \sqrt{\lambda}\,\pi + B \cos \sqrt{\lambda}\,\pi = 0. \end{aligned}$$

A non–trivial solution emerges only when $A = 0$ and $\cos \sqrt{\lambda}\,\pi = 0$. The latter condition gives

$$\sqrt{\lambda_n} = \frac{1}{2}, \frac{3}{2}, \frac{5}{2}, \ldots \qquad \longrightarrow \qquad \sqrt{\lambda_n} = \frac{2n-1}{2}, \quad n = 1, 2, 3, \ldots.$$

The solutions of the boundary value problem are therefore provided by the functions

$$y_n(x) = k_n \cos(n - 1/2)\, x, \quad n = 1, 2, 3, \ldots$$

where k_n are arbitrary constants.   □

Remarks 9.2   In Examples 9.1 and 9.2 we recognize again certain features that were apparent in the motivational problem of the elastic string in Section 9.1.

1. A non–trivial solution of the boundary value problem exists only for specific values of the parameter λ. These are the *eigenvalues*.

2. The solution $y_n(x)$ of the boundary value problem corresponding to a specific eigenvalue λ_n is called the *eigenfunction*.

3. A solution $y_n(x)$ is not unique but can be multiplied by an arbitrary constant and still remains a solution.

4. The non–uniqueness of solutions permits a suitable choice of the arbitrary constant. This procedure will be discussed in more detail in Section 9.4. For the solution of Example 9.2, for illustration, we can calculate

$$\int_0^\pi y_n^2(x)\,dx = k_n^2 \int_0^\pi \cos^2\left(n - 1/2\right) x\,dx$$

$$= \frac{k_n^2}{2}\left[x + \frac{\sin\left(n - 1/2\right) x\ \cos\left(n - 1/2\right) x}{n - 1/2} \right]_0^\pi = \frac{k_n^2 \pi}{2}$$

and require that the integral equals 1. The choice $k_n = \sqrt{2/\pi}$ therefore provides the so–called *normalized eigenfunctions*

$$y_n(x) = \sqrt{\frac{2}{\pi}}\,\cos\left(n - 1/2\right) x\,. \qquad \triangle$$

9.3 STURM BOUNDARY VALUE PROBLEMS

Let $\hat{L}$ be the linear differential operator whose action on the function y is defined by

$$\hat{L}\,y(x)\ =\ \left[p(x)y'(x) \right]' + q(x)y(x)\,. \qquad (9.18)$$

The boundary value problem, defined by

$$\hat{L}\,y(x) = h(x)\,, \qquad a < x < b\,, \qquad (9.19)$$

$$\hat{R}_1 y = \alpha_1 y(a) + \alpha_2 y'(a) = \eta_1\,, \qquad (9.20)$$

$$\hat{R}_2 y = \beta_1 y(b) + \beta_2 y'(b) = \eta_2\,, \qquad (9.21)$$

is associated with the name of the mathematician Sturm. In (9.20) and (9.21), η_1 and η_2 have known values, $\alpha_1^2 + \alpha_2^2 > 0$ and $\beta_1^2 + \beta_2^2 > 0$ while $p(x)$ is a non–negative continuously differentiable function and $q(x)$ is a continuous function for $a < x < b$.

Remarks 9.3   Every differential equation of second order expressed in the standard form $y'' + a_1(x)y' + a_2(x)y = g(x)$ can be written in the form (9.19). To see this, multiply the differential equation by $p(x) = e^{\int a_1(x)dx}$. It becomes clear then that $p'(x) = a_1(x)e^{\int a_1(x)dx} = a_1(x)p(x)$ and thus $py'' + a_1 py' + a_2 py = pg$ which can be rewritten as $(py')' + qy = h$ with the identifications $q = a_2 p$ and $h = pg$. $\qquad \triangle$

Now let $y_1(x), y_2(x)$ be a fundamental set of solutions of the homogeneous version of (9.19), that is, y_1 and y_2 fulfil the boundary conditions and satisfy $\hat{L}y_1 = 0, \hat{L}y_2 = 0$. The general solution of the inhomogeneous boundary value problem (9.19)–(9.21) may be written in the usual form

$$y(x) = Ay_1(x) + By_2(x) + y_p(x) \qquad (9.22)$$

with a particular integral $y_p(x)$. The integration constants A and B are chosen so that

$$\hat{R}_1 y = A\hat{R}_1 y_1 + B\hat{R}_1 y_2 + \hat{R}_1 y_p = \eta_1 \qquad (9.23)$$
$$\hat{R}_2 y = A\hat{R}_2 y_1 + B\hat{R}_2 y_2 + \hat{R}_2 y_p = \eta_2 \ . \qquad (9.24)$$

These equations for A and B may be re–expressed in the matrix form

$$\begin{bmatrix} \hat{R}_1 y_1 & \hat{R}_1 y_2 \\ \hat{R}_2 y_1 & \hat{R}_2 y_2 \end{bmatrix} \begin{bmatrix} A \\ B \end{bmatrix} = \begin{bmatrix} \eta_1 - \hat{R}_1 y_p \\ \eta_2 - \hat{R}_2 y_p \end{bmatrix} \qquad (9.25)$$

and so the determination of A and B requires the existence and uniqueness of the solution of a pair of simultaneous linear equations. Let D be the value of the determinant of the matrix on the left–hand side of (9.25), i.e.,

$$D = \begin{vmatrix} \hat{R}_1 y_1 & \hat{R}_1 y_2 \\ \hat{R}_2 y_1 & \hat{R}_2 y_2 \end{vmatrix} . \qquad (9.26)$$

A necessary and sufficient condition for (9.25) to have a unique solution is that $D \neq 0$. In this case, the inhomogeneous problem (9.19)–(9.21) has a unique solution and the corresponding homogeneous problem has only the trivial solution.

Alternatively, if $D = 0$, the inhomogeneous problem does not have a unique solution whereas the corresponding homogeneous problem can be solved — but its solution contains an arbitrary constant.

Example 9.3   Solve the differential equation $y'' + y = 1$ with the boundary conditions $\hat{R}_1 y = y(0) + y'(0) = 0$ and $\hat{R}_2 y = y(\pi) = 0$.

Solution 9.3   The functions $y_1(x) = \cos x$ and $y_2(x) = \sin x$ are a fundamental set of solutions. Thus

$$\begin{vmatrix} \hat{R}_1 y_1 & \hat{R}_1 y_2 \\ \hat{R}_2 y_1 & \hat{R}_2 y_2 \end{vmatrix} = \begin{vmatrix} 1 & 1 \\ -1 & 0 \end{vmatrix} = 1 \ .$$

Therefore, the inhomogeneous problem has a unique solution. Now

$$y(x) = ay_1(x) + by_2(x) + y_p(x) = a \cos x + b \sin x + 1$$

where the particular integral $y_p = 1$ is obtained by guessing. The boundary conditions require

$$
\begin{aligned}
1 + a + b &= 0 \\
1 - a &= 0
\end{aligned}
\qquad \longrightarrow \qquad
\begin{aligned}
a &= 1 \\
b &= -2
\end{aligned}
$$

and the required solution is therefore

$$
y = \cos x - 2 \sin x + 1 .
$$

It is checked easily that $y \equiv 0$ is the only solution of the homogeneous problem.

Consider now the differential equation $y'' + y = 1$ as before but with the new boundary conditions

$$
\hat{R}_1 y = y(0) = 0 , \qquad \hat{R}_2 y = y(\pi) = 0 .
$$

In this case,

$$
\begin{vmatrix} \hat{R}_1 y_1 & \hat{R}_1 y_2 \\ \hat{R}_2 y_1 & \hat{R}_2 y_2 \end{vmatrix} = \begin{vmatrix} 1 & 0 \\ -1 & 0 \end{vmatrix} = 0
$$

and so the inhomogeneous problem has no solution whereas the homogeneous problem has the solution $y = C \sin x$ in which C is an arbitrary constant. □

9.4   THE STURM–LIOUVILLE EIGENVALUE PROBLEM

The differential equation

$$
\hat{L} y(x) + \lambda r(x) y(x) = 0 , \qquad a < x < b \tag{9.27}
$$

where $\hat{L}$ is defined by (9.18) and the boundary conditions

$$
\hat{R}_1 y = 0 , \qquad \hat{R}_2 y = 0 \tag{9.28}
$$

form a **Sturm–Liouville eigenvalue problem**. The differential equation contains the parameter λ (the eigenvalue) and we assume that p, p', q and r are continuous and $p > 0$ and $r > 0$ for $a < x < b$.

Before proceeding further, the **Lagrange identity** is derived. Let $u_1(x)$ and $u_2(x)$ be two functions with continuous second derivatives for $a < x < b$. The Lagrange identity states that

$$
\begin{aligned}
\int_a^b \left[u_2(\hat{L} u_1) - u_1(\hat{L} u_2) \right] \, dx &= \int_a^b \left[(pu_1')' u_2 - u_1(pu_2')' \right] \, dx \\
&= \left[p(u_1' u_2 - u_1 u_2') \right]_a^b .
\end{aligned}
\tag{9.29}
$$

The result was obtained from the definition (9.18) of the operator $\hat{L}$ and using integrating by parts. In a further specialization, the right–hand side of (9.29) vanishes if u_1 and u_2 fulfil the boundary conditions $\hat{R}_i u_j = 0$ $(i, j = 1, 2)$.

Various *theorems* can be proved rigorously in the theory of Sturm–Liouville eigenvalue problems. Here it suffices to state results and provide some guidance as to how these results are proved.

1. All the eigenvalues of the Sturm–Liouville eigenvalue problem (9.27), (9.28) are **real–valued**. The proof is based on the observation that

$$(\lambda - \lambda^*) \int_a^b r(x)u(x)u^*(x)\, dx + \int_a^b \left[u^*(\hat{L}\, u) - u(\hat{L}\, u^*) \right] dx = 0$$

where the superscript * indicates the complex conjugate. The Lagrange identity, in combination with the boundary conditions, is now used to establish that the second of these integrals vanishes. Therefore $\lambda = \lambda^*$ since the first integral is inherently positive.

2. All the eigenvalues are **simple**, that is, each eigenvalue has an eigenfunction that is unique up to a constant scaling factor. The proof is established by a contradiction argument.

3. The eigenvalues may be ordered into an unbounded sequence

$$\lambda_1 < \lambda_2 < \lambda_3 < \ldots < \lambda_n < \ldots . \tag{9.30}$$

Given any $M > 0$, there is an integer n such that $\lambda_k > M$ for all $k > n$.

4. As the eigenvalues form an infinite sequence, there is consequently an infinite sequence of eigenfunctions frequently denoted by $\phi_n(x)$.

5. Two eigenfunctions $\phi_j(x)$ and $\phi_k(x)$ corresponding to distinct eigenvalues λ_j and λ_k respectively of the Sturm–Liouville eigenvalue problem are **orthogonal** in the sense that

$$\int_a^b r(x)\phi_j(x)\phi_k(x)dx = 0, \qquad j \neq k. \tag{9.31}$$

The proof uses the Lagrange identity (9.29) with u_1 identified with ϕ_j and u_2 identified with ϕ_k. By definition, $\hat{L}\,\phi_j + \lambda_j r\phi_j = 0$ and $\hat{L}\,\phi_k + \lambda_k r\phi_k = 0$.

6. The eigenfunctions ϕ_n can be normalized by the requirement that

$$\int_a^b r(x)\phi_n^2(x)dx = 1, \qquad n = 1, 2, 3, \ldots . \tag{9.32}$$

Based on (9.31) and (9.32), the eigenfunctions $\phi_n(x)$, $n = 1, 2, 3, \ldots$ are said to form an **orthonormal set**.

Let $y(x)$ be a function that obeys the two boundary conditions (9.28), $\hat{R}_1 y = \hat{R}_2 y = 0$ and assume that y and y' are piecewise continuous, then $y(x)$ can be expanded in terms of the normalized eigenfunctions $\phi_n(x)$ in the form

$$y(x) = \sum_{n=1}^{\infty} a_n \phi_n(x), \qquad a_n = \int_a^b r(x)\phi_n(x)y(x)\,dx. \qquad (9.33)$$

The determination of a_n uses the orthonormal properties of the set of eigenfunctions $\phi_n(x)$ and follows from the calculation

$$\int_a^b r(x)\phi_n(x)\phi(x)\,dx = \int_a^b r(x)\phi_n(x)\sum_{k=1}^{\infty} a_k \phi_k(x)\,dx$$

$$= \sum_{k=1}^{\infty} a_k \int_a^b r(x)\phi_k(x)\phi_n(x)\,dx = a_n.$$

The only non–zero integral in this summation occurs when $n = k$: all other integrals in the summation vanish due to the orthogonal properties of the set of eigenfunctions $\phi_n(x)$.

Now is a good time to reassess the introductory problem — the elastic string. After separation of variables, it was found in (9.7) that

$$F''(x) + \alpha F(x) = 0, \quad F(0) = F(l) = 0.$$

Clearly $\alpha_n = (n\pi/l)^2$ are the *eigenvalues* and $F_n(x) = \sin(n\pi x/l)$ are the corresponding *eigenfunctions*. The *normalized* eigenfunctions are given by $\phi_n(x) = A_n F_n(x)$ where A_n is to be found. In terms of a Sturm–Liouville eigenvalue problem, the string problem corresponds to the special case $r(x) = 1$. Thus A_n, the normalizing coefficient, is obtained from the condition

$$\int_0^l r(x)\phi_n(x)\phi_n(x)\,dx = A_n^2 \int_0^l \sin^2 \frac{n\pi x}{l}\,dx = 1. \qquad (9.34)$$

Using the trigonometric identity $2\sin^2\theta = 1 - \cos 2\theta$, standard methods of integration yield

$$1 = \frac{A_n^2}{2}\int_0^l \left(1 - \cos \frac{2n\pi x}{l}\right) dx = \frac{A_n^2 l}{2} \qquad (9.35)$$

from which it follows that $A_n = \sqrt{2/l}$. Therefore

$$\phi_n(x) = \sqrt{\frac{2}{l}} \sin \frac{n\pi x}{l}, \qquad n = 1, 2, 3, \ldots \qquad (9.36)$$

form a normalized set of eigenfunctions for the string problem. A function $y(x)$ satisfying $y(0) = y(l) = 0$ can now be expanded into

$$y(x) = \sum_{n=1}^{\infty} a_n \phi_n(x) = \sqrt{\frac{2}{l}} \sum_{n=1}^{\infty} a_n \sin \frac{n\pi x}{l}, \qquad (9.37)$$

$$a_n = \sqrt{\frac{2}{l}} \int_0^l y(x) \sin \frac{n\pi x}{l} \, dx. \qquad (9.38)$$

The series (9.37) converges everywhere if $y(x)$ is a continuous function. Expansions of the type (9.37), (9.38) are closely related to the concept of Fourier series.

Example 9.4   Find the eigenvalues and eigenfunctions of the boundary value problem $y'' + \lambda y = 0$, $y(0) = 0$, $y'(1) + y(1) = 0$.

Solution 9.4   The character of the equation depends on the value of λ. There are four different cases to be considered.

1. $\lambda = 0$. It follows immediately that $y(x) = c_1 x + c_2$. Applying the boundary conditions leads to $c_2 = 0$ and $2c_1 + c_2 = 0$ which can only be fulfilled for $c_1 = c_2 = 0$ and therefore only the trivial solution exists.

2. $\lambda > 0$. In this case

$$y(x) = c_1 \sin \sqrt{\lambda}\, x + c_2 \cos \sqrt{\lambda}\, x$$
$$\longrightarrow \quad y'(x) = c_1 \sqrt{\lambda} \cos \sqrt{\lambda}\, x - c_2 \sqrt{\lambda} \sin \sqrt{\lambda}\, x.$$

The first boundary condition requires that $0 = c_2$. Taking this into account, the second boundary condition reduces to

$$c_1 \sqrt{\lambda} \cos \sqrt{\lambda} + c_1 \sin \sqrt{\lambda} = 0.$$

Since $c_2 = 0$ then c_1 must be non-zero for a non-trivial solution. Thus

$$\sqrt{\lambda} \cos \sqrt{\lambda} + \sin \sqrt{\lambda} = 0 \qquad \longrightarrow \qquad \sqrt{\lambda} = -\tan \sqrt{\lambda}.$$

This is a **transcendental equation** and can only be solved approximately using numerical or graphical means. From Figure 9.2, it can be seen that

$$\sqrt{\lambda_1} \approx 2.0 \qquad \longrightarrow \qquad \lambda_1 \approx 4.0$$
$$\sqrt{\lambda_2} \approx 4.5 \qquad \longrightarrow \qquad \lambda_2 \approx 20.3$$

$$\vdots \qquad\qquad \vdots \qquad \vdots$$

$$\sqrt{\lambda_n} \approx (n - \tfrac{1}{2})\pi \qquad \longrightarrow \qquad \lambda_n \approx (n - \tfrac{1}{2})^2 \pi^2, \quad n = 3, 4, \ldots.$$

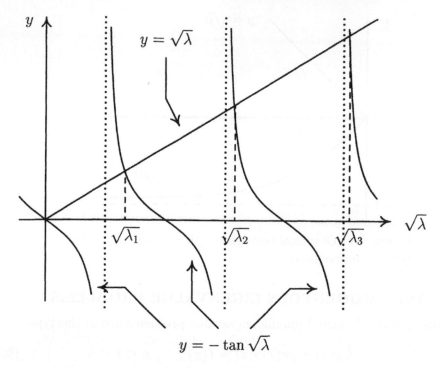

Figure 9.2: Graphical solution of $\sqrt{\lambda} = -\tan\sqrt{\lambda}$ with the first three non–trivial roots, $\sqrt{\lambda_1}$, $\sqrt{\lambda_2}$, $\sqrt{\lambda_3}$ indicated.

The unnormalized eigenfunctions are

$$y_n(x) = k_n \sin\sqrt{\lambda_n}\,x\,, \qquad n = 1, 2, 3, \ldots\,.$$

3. $\lambda < 0$. Let $\mu = -\lambda$ then $\mu > 0$ and

$$y(x) = c_1 \sinh\sqrt{\mu}\,x + c_2 \cosh\sqrt{\mu}\,x\,.$$

Again, the first boundary condition requires $c_2 = 0$ while the second boundary condition leads to the relation

$$\sqrt{\mu} = -\tanh\sqrt{\mu}\,.$$

Figure 9.3 shows that no eigenvalues exist for $\mu > 0$ and thus only the trivial solution remains.

4. If λ is complex–valued, the trivial solution $y \equiv 0$ is the only possible solution because this boundary value problem is a Sturm–Liouville eigenvalue problem in which all eigenvalues must be real–valued.   □

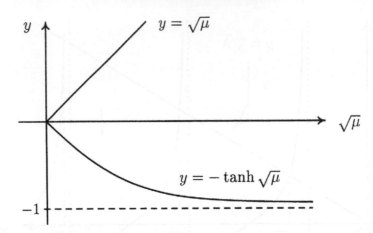

Figure 9.3: Graphical solution of $\sqrt{\mu} = -\tanh\sqrt{\mu}$. Only the trivial solution exists.

9.5 INHOMOGENEOUS EIGENVALUE PROBLEMS

Inhomogeneous Sturm–Liouville eigenvalue problems are of the type

$$\hat{L}\,y(x) + \mu r(x)y(x) = f(x)\,, \qquad a < x < b \qquad (9.39)$$

where the differential operator $\hat{L}$ is again given by (9.18) with the boundary conditions

$$\hat{R}_1 y = 0\,, \quad \hat{R}_2 y = 0\,. \qquad (9.40)$$

Our aim here is to obtain a solution of (9.39), (9.40) in terms of an **eigenfunction expansion**. For that purpose, we first solve the homogeneous Sturm–Liouville eigenvalue problem and obtain the eigenvalues λ_n and the normalized eigenfunctions $\phi_n(x)$ such that

$$\hat{L}\,\phi_n(x) + \lambda_n r(x)\phi_n(x) = 0\,, \qquad \hat{R}_1\phi_n(x) = \hat{R}_2\phi_n(x) = 0 \qquad (9.41)$$

with

$$\int_a^b r(x)\phi_n(x)\phi_m(x)\,dx = \begin{cases} 0 & m \neq n \\ 1 & m = n\,. \end{cases} \qquad (9.42)$$

Next, we multiply both sides of (9.39) by $\phi_n(x)$ and integrate from a to b to obtain

$$\int_a^b \left[[\hat{L}\,y(x)]\,\phi_n(x) + \mu r(x)y(x)\phi_n(x) \right] dx = \int_a^b f(x)\phi_n(x)\,dx\,. \qquad (9.43)$$

Using the definition (9.18) of $\hat{L}$ we obtain by using integration by parts

$$\int_a^b f\phi_n\,dx = \int_a^b \left[(py')'\,\phi_n + qy\phi_n + \mu ry\phi_n \right] dx$$

$$= \left[py'\phi_n \right]_a^b + \int_a^b \left[(-py'\phi_n') + qy\phi_n + \mu r y\phi_n \right] dx$$

$$= \left[p \left(y'\phi_n - y\phi_n' \right) \right]_a^b + \int_a^b y \left[(p\phi_n')' + q\phi_n + \mu r\phi_n \right] dx .$$

Using the fact that the eigenfunctions ϕ_n satisfy (9.41) and y satisfies the boundary conditions (9.40), the last relation reduces to

$$(\mu - \lambda_n) \int_a^b r(x) y(x) \phi_n(x) \, dx = \int_a^b f(x) \phi_n(x) dx . \tag{9.44}$$

Suppose now that

$$y(x) = \sum_{n=1}^{\infty} b_n \phi_n(x) \tag{9.45}$$

is the expansion of $y(x)$ in terms of the eigenfunctions $\phi_n(x)$. Substitution of (9.45) into (9.44) yields [upon exploitation of the orthonormality relations (9.42)]

$$(\mu - \lambda_n) \, b_n = c_n \tag{9.46}$$

where we have defined

$$c_n = \int_a^b f(x) \phi_n(x) \, dx . \tag{9.47}$$

We therefore obtain the particular integral in its final form as per

$$y(x) = \sum_{n=1}^{\infty} \frac{c_n}{\mu - \lambda_n} \phi_n(x) . \tag{9.48}$$

It is clear that the solution (9.48) is valid only if μ is not an eigenvalue of the homogeneous Sturm–Liouville eigenvalue problem, that is, $\mu \neq \lambda_n$, $n = 1, 2, 3, \dots$.

If, however, $\mu = \lambda_k$ for some integer k then (9.46) requires that

$$c_k = \int_a^b f(x) \phi_k(x) \, dx = 0 . \tag{9.49}$$

This last relation is a consistency condition and is called the **orthogonality condition**. As a consequence of this requirement, the solution is no longer unique because the coefficient b_k in the expansion (9.48) of $y(x)$ remains arbitrary.

Summary: The inhomogeneous Sturm–Liouville eigenvalue problem defined by (9.39), (9.40) has a unique solution for every continuous function $f(x)$ provided μ is not an eigenvalue of the corresponding homogeneous Sturm–Liouville eigenvalue problem. The solution (9.48) converges uniformly for every continuous function $f(x)$.

Example 9.5   Solve the inhomogeneous boundary value problem

$$y'' + 2y = -x + \frac{2x^2}{3}, \qquad y(0) = 0, \quad y(1) + y'(1) = 0 \tag{9.50}$$

as an eigenfunction expansion in terms of the eigenfunctions of the corresponding homogeneous Sturm–Liouville eigenvalue problem and by a direct method. Compare the two solutions.

Solution 9.5   In order to find the desired expansion for $y(x)$, it is first necessary to find the eigenfunctions $\phi_n(x)$ of the corresponding Sturm–Liouville eigenvalue problem

$$\phi_n''(x) + \lambda_n \phi_n(x) = 0, \qquad \phi_n(0) = 0, \quad \phi_n(1) + \phi_n'(1) = 0.$$

The normalized eigenfunctions corresponding to eigenvalues λ_n are (compare with Example 9.4)

$$\phi_n(x) = A_n \sin \sqrt{\lambda_n} x, \qquad \sqrt{\lambda_n} = -\tan \sqrt{\lambda_n},$$

where the normalization constant is $A_n = \sqrt{2/(1 + \cos^2 \sqrt{\lambda_n})}$. With $\mu = 2$, we have from (9.48)

$$y(x) = \sum_{n=1}^{\infty} \frac{c_n}{2 - \lambda_n} \phi_n(x)$$

where

$$c_n = \int_0^1 f(x)\phi_n(x)\,dx = \int_0^1 \left(-x + \frac{2x^2}{3}\right) A_n \sin \sqrt{\lambda_n} x\,dx = A_n\left(-I_1 + \frac{2I_2}{3}\right).$$

The two integrals I_1 and I_2 can be evaluated as

$$I_1 = \int_0^1 x \sin \sqrt{\lambda_n} x\,dx = \left[\frac{\sin \sqrt{\lambda_n} x}{\lambda_n} - \frac{x \cos \sqrt{\lambda_n} x}{\sqrt{\lambda_n}}\right]_0^1$$

$$= \frac{\sin \sqrt{\lambda_n}}{\lambda_n} - \frac{\cos \sqrt{\lambda_n}}{\sqrt{\lambda_n}} = -\frac{2 \cos \sqrt{\lambda_n}}{\sqrt{\lambda_n}},$$

the last step arising by substituting $\sqrt{\lambda_n} = -\tan \sqrt{\lambda_n}$. Further

$$I_2 = \int_0^1 x^2 \sin \sqrt{\lambda_n} x\,dx$$

$$= \left[\frac{2x}{\lambda_n} \sin \sqrt{\lambda_n} x - \left(\frac{x^2}{\sqrt{\lambda_n}} - \frac{2}{\lambda_n \sqrt{\lambda_n}}\right) \cos \sqrt{\lambda_n} x\right]_0^1$$

$$= \frac{2\sin\sqrt{\lambda_n}}{\lambda_n} - \left(\frac{1}{\sqrt{\lambda_n}} - \frac{2}{\lambda_n\sqrt{\lambda_n}}\right)\cos\sqrt{\lambda_n} - \frac{2}{\lambda_n\sqrt{\lambda_n}}$$

$$= \frac{1}{\lambda_n\sqrt{\lambda_n}}\left[2\sqrt{\lambda_n}\sin\sqrt{\lambda_n} - (\lambda_n-2)\cos\sqrt{\lambda_n} - 2\right]$$

$$= \frac{1}{\lambda_n\sqrt{\lambda_n}}\left[(2-3\lambda_n)\cos\sqrt{\lambda_n} - 2\right].$$

Putting the two terms together gives

$$c_n = A_n\left\{\frac{2\cos\sqrt{\lambda_n}}{\sqrt{\lambda_n}} + \frac{2}{3\sqrt{\lambda_n}\lambda_n}\left[(2-3\lambda_n)\cos\sqrt{\lambda_n} - 2\right]\right\}$$

$$= \frac{2A_n}{3\lambda_n\sqrt{\lambda_n}}\left[3\lambda_n\cos\sqrt{\lambda_n} + (2-3\lambda_n)\cos\sqrt{\lambda_n} - 2\right]$$

$$= \frac{4A_n}{3\lambda_n\sqrt{\lambda_n}}\left(\cos\sqrt{\lambda_n} - 1\right).$$

The solution is

$$y(x) = \sum_{n=1}^{\infty}\frac{c_n}{2-\lambda_n}\phi_n(x) = \sum_{n=1}^{\infty}\frac{4(1-\cos\sqrt{\lambda_n})}{3\lambda_n\sqrt{\lambda_n}(\lambda_n-2)}A_n\phi_n(x)$$

$$= \frac{8}{3}\sum_{n=1}^{\infty}\frac{(1-\cos\sqrt{\lambda_n})}{\lambda_n\sqrt{\lambda_n}(\lambda_n-2)(1+\cos^2\sqrt{\lambda_n})}\sin\sqrt{\lambda_n}x. \qquad (9.51)$$

Alternatively, this inhomogeneous boundary value problem can be solved directly by determining the complementary function and a particular integral. The complementary function is $y_c(x) = A\cos\sqrt{2}x + B\sin\sqrt{2}x$. By calculating

$$\begin{vmatrix} \hat{R}_1y_1 & \hat{R}_1y_2 \\ \hat{R}_2y_1 & \hat{R}_2y_2 \end{vmatrix} = \sin\sqrt{2} + \sqrt{2}\cos\sqrt{2} \neq 0,$$

the inhomogeneous problem is seen to have a unique solution. The general solution is therefore

$$y(x) = A\cos\sqrt{2}x + B\sin\sqrt{2}x + y_p(x)$$

where the particular integral y_p can be obtained (by using the method of undetermined coefficients) in the form $y_p = a + bx + cx^2$. It is found that $a = -1/3$, $b = -1/2$, $c = 1/3$ and so the general solution is

$$y = A\cos\sqrt{2}x + B\sin\sqrt{2}x - \frac{1}{3} - \frac{x}{2} + \frac{x^2}{3}.$$

The integration constants A and B follow from the requirement that the particular solution must satisfy the boundary conditions. We obtain

$$A = \frac{1}{3}, \qquad B = \frac{1 + \sqrt{2}\sin\sqrt{2} - \cos\sqrt{2}}{3(\sin\sqrt{2} + \sqrt{2}\cos\sqrt{2})}.$$

Finally,

$$y(x) = \frac{1}{3}\cos\sqrt{2}x + \frac{1 + \sqrt{2}\sin\sqrt{2} - \cos\sqrt{2}}{3(\sin\sqrt{2} + \sqrt{2}\cos\sqrt{2})}\sin\sqrt{2}x - \frac{1}{3} - \frac{x}{2} + \frac{x^2}{3}. \quad (9.52)$$

The two solutions (9.51) and (9.52) appear very different from each other: (9.51) is an infinite series whereas (9.52) is a finite expression involving a polynomial of order two and two trigonometric terms. Both solutions are equally valid and represent the same function; the infinite series (9.51) converges uniformly for all values $0 \le x \le 1$. $\square$

Remarks 9.4   This section closes with a few stricter observations about the mathematical nature of eigenfunction expansions. Of particular interest are the conditions under which a function has an eigenfunction expansion, and the convergence properties of such an expansion when it exists. An important theorem in the theory of Sturm–Liouville eigenvalue problems is:

- A function $f(x)$ which is piecewise continuous and which has a piecewise continuous derivative can be expanded in terms of the eigenfunctions of the Sturm–Liouville eigenvalue problem (9.27) and (9.28) over a given interval $a \le x \le b$. At each point of the open interval $a < x < b$ the expansion (with coefficients calculated as described earlier) converges to

$$\frac{1}{2}\lim_{h\to 0}\left[f(x+h) + f(x-h)\right].$$

This theorem identifies an important point in the theory of eigenfunction expansions that arises when a function $y(x)$ is expanded in terms of the normalized eigenfunctions ϕ_n of a Sturm–Liouville eigenvalue problem in which $y(x)$ and $\phi_n(x)$ fulfil *different* boundary conditions. Formally, one can still write

$$y(x) = \sum_{n=1}^{\infty} b_n\phi_n(x), \qquad b_n = \int_a^b r(x)y(x)\,\phi_n(x)\,dx$$

and the series will converge at all points of the interval $a < x < b$ for a continuous function $y(x)$. At the boundary points $x = a$ and $x = b$, however, it is clearly impossible to achieve convergence if y and the eigenfunctions satisfy different boundary conditions.

To provide evidence of this difficulty, consider the expansion of $y(x) = x$ in terms of the normalized eigenfunctions $\phi_n(x) = \sqrt{2}\sin(n\pi x)$ of $\phi'' + \lambda\phi = 0$ with $\phi(0) = \phi(1) = 0$. The formal result is

$$x = \frac{2}{\pi} \sum_{n=1}^{\infty} (-1)^{(n+1)} \frac{\sin(n\pi x)}{n}. \tag{9.53}$$

Convergence of the eigenfunction expansion (9.53) can be expected for all values of $x \in [0,1)$ because at the boundary point $x = 0$ the function y and all the eigenfunctions ϕ_n are zero. At the boundary point $x = 1$ however, the expansion of y in terms of ϕ_n forces an incorrect boundary value on y; the closer the independent variable x gets to 1, the more will the value of y, as calculated from the expansion, deviate from its actual value. Thus, near $x = 1$ the partial sums of (9.53) provide a very poor representation of y.

Figure 9.4 illustrates the graph of $y(x) = x$ together with the approximation to x derived by truncating the eigenvalue expansion (9.53) after 1 and 10 terms (in the left part of the figure) and 1 and 50 terms (in the right part of the figure). This type of fluctuating behaviour is known as the **Gibbs phenomenon**. At a point of discontinuity, the oscillations accompanying the Gibbs phenomenon have an overshoot of approximately 18% of the amplitude of the discontinuity. As n, the number of eigenfunctions included in

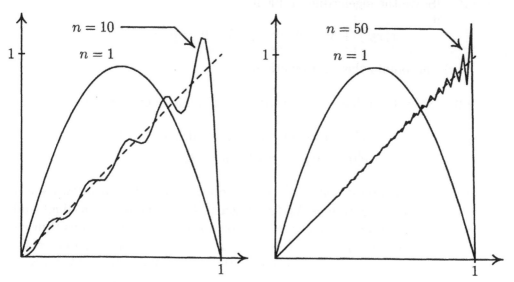

Figure 9.4: Graphs of $y(x) = x$ (dashed line) compared with approximations of the infinite sum truncated after 1, 10, 50 terms.

the series is increased, the non–uniform nature of the convergence ensures that the Gibbs oscillations are merely telescoped towards the discontinuity at a rate proportional to $1/n$, but never vanish.   △

9.6   TUTORIAL EXAMPLES 9

T 9.1   Find the eigenvalues and eigenfunctions of the boundary value problem $y''(x) + \lambda y(x) = 0$, $y(0) = 0$, $y'(1) = 0$.

T 9.2   Find the eigenvalues and eigenfunctions of the boundary value problem $y''(x) + \lambda y(x) = 0$, $y'(0) = 0$, $y'(1) = 0$.

T 9.3   Find eigenvalues and eigenfunctions of the boundary value problem

$$y''(x) + \lambda y(x) = 0, \qquad y'(0) = 0, \quad y(1) + y'(1) = 0.$$

Establish if $\lambda = 0$ is an eigenvalue. Find an approximate value for the eigenvalue of smallest value. Estimate λ_n for large values of n.

T 9.4   Determine real–valued eigenvalues of the boundary value problem

$$y''(x) + (\lambda + 1)y'(x) + \lambda y(x) = 0, \qquad y'(0) = 0, \quad y(1) = 0,$$

if any exist, and the form of the corresponding eigenfunction(s).

T 9.5   Solve the eigenvalue problem

$$x(xy')' + \lambda y = 0, \qquad y'(1) = 0, \quad y'(e^{2\pi}) = 0$$

for $y(x)$ and obtain the eigenfunctions.

T 9.6   Solve the inhomogeneous Sturm–Liouville boundary value problem

$$y''(x) + 9y(x) = \cos x, \qquad 0 < x < \pi/4,$$
$$y'(0) = 0, \quad y(\pi/4) + y'(\pi/4) = 0.$$

Does it have a unique solution? If yes, obtain the solution in two ways. Firstly, find the eigenvalues λ_n (establish if $\lambda = 0$ is an eigenvalue) and normalized eigenfunctions $\phi_n(x)$ of the corresponding homogeneous Sturm–Liouville eigenvalue problem. Expand $y(x)$ in terms of the eigenfunctions $\phi_n(x)$. Secondly, solve the problem directly and compare the two solutions.

10

Calculus of Variations

10.1 INTRODUCTION

Variable quantities called **functionals** play an important role in many problems of mathematics. A functional is a relationship which assigns a number to each function (or curve) belonging to some function class (for example, the space of continuously differentiable functions in the interval $a < x < b$). In effect, a functional is a relation in which the variable is a function. This idea is first illustrated with some historical applications.

10.2 HISTORICAL PROBLEMS

The Brachistochrone

This problem was originally posed by Johann Bernoulli in the late 17th century as a challenge to contemporary mathematicians and particularly his brother Jakob. He asked the question: given two *fixed* points $A(x_A, y_A)$ and $B(x_B, y_B)$ in a vertical plane, along what continuously differentiable curve joining A and B in the plane must a particle of mass m move under constant gravity to minimize the transit time? The geometry of this problem is shown in Figure 10.1.

Let $y(x)$ be any continuously differentiable curve joining A and B and let P be the point (x, y), then conservation of energy yields

$$\frac{1}{2} mv^2 + mgy = \frac{1}{2} mv_A^2 + mgy_A \qquad (10.1)$$

where v is the particle speed at P, v_A is the initial speed of the particle at A and g is the gravitational acceleration. Hence

$$v = \sqrt{v_A^2 + 2g(y_A - y)}. \qquad (10.2)$$

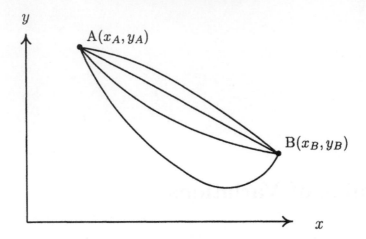

Figure 10.1: The Brachistochrone problem. Paths from A to B
with constant force of gravity in the negative y direction.

In terms of the *arc length* of this curve,

$$v = \frac{ds}{dt} = \frac{ds}{dx}\frac{dx}{dt} = \sqrt{1 + y'^2}\,\frac{dx}{dt} \tag{10.3}$$

and hence

$$\frac{dx}{dt} = \sqrt{\frac{v_A^2 + 2g(y_A - y)}{1 + y'^2}}. \tag{10.4}$$

Let τ be the transit time from A to B, then we have

$$\tau(y) = \int_{x_A}^{x_B} \sqrt{\frac{1 + y'^2}{v_A^2 + 2g(y_A - y)}}\,dx. \tag{10.5}$$

The transit time is thus a functional of $y(x)$ and should be minimized subject
to the conditions $y(x_A) = y_A$, $y(x_B) = y_B$. In the classical problem, the
particle is released from rest and the solution is part of a **cycloid**.

The hanging chain

The problem is to find the shape of an inextensible chain of *fixed* length l
suspended between two *fixed* points $A(x_A, y_A)$ and $B(x_B, y_B)$ in a constant
gravitational field as shown in Figure 10.2. The idea is that the adopted
configuration of the chain minimizes the chain's potential energy.

The **potential energy** of a section of the chain of length ds at point
(x, y) is $\rho g y\, ds$ where $\rho(x)$ is the linear density of the chain. Hence the
potential energy of the entire chain is

$$V(y) = \int_0^l \rho g y\, ds = \int_{x_A}^{x_B} \rho g y \sqrt{1 + y'^2}\, dx \tag{10.6}$$

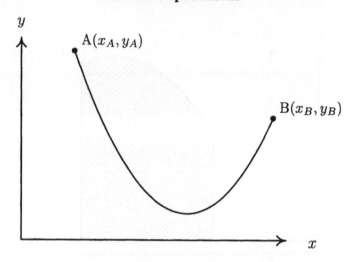

Figure 10.2: The hanging chain.

where $y(x)$ must also satisfy the length constraint

$$l = \int_{x_A}^{x_B} \sqrt{1 + y'^2}\, dx \,. \tag{10.7}$$

The potential energy V is therefore a functional of y and should be minimized subject to the length constraint. Chains of uniform linear density — for example, power cables between pylons — adopt the **catenary** shape.

The isoperimetric problem

This problem was first posed by the ancient Greeks and involves finding the curve of given length l which encloses maximum area and passes through two *fixed points* $A(x_A, y_A)$ and $B(x_B, y_B)$, see Figure 10.3.

The task is find a function $y(x)$ that maximizes

$$A(y) = \int_{x_A}^{x_B} y\, dx \tag{10.8}$$

subject to the constraint

$$l = \int_{x_A}^{x_B} \sqrt{1 + y'^2}\, dx \,. \tag{10.9}$$

The area is again a constrained functional of y.

A variant of this problem occurs when the point B is not fixed but is allowed to travel along a given curve $g(x)$. This gives rise to a **transversality condition** for the location of B. Such problems will not be considered here.

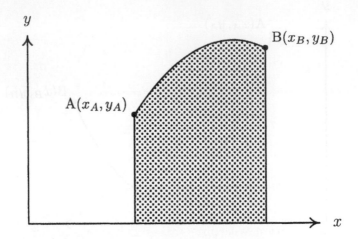

Figure 10.3: The isoperimetric problem.

The geodesic problem

This is the problem of finding the shortest distance between two *fixed* points in a region (for example, on a surface or through space–time) given a measure or *metric* indicating how distance should be measured between parametrically adjacent points. The simplest geodesic problem is to find the shortest plane curve between two fixed points $A(x_A, y_A)$ and $B(x_B, y_B)$, that is, minimize

$$S(y) = \int_{x_A}^{x_B} \sqrt{1 + y'^2}\, dx \tag{10.10}$$

where $y(x_A) = y_A$ and $y(x_B) = y_B$. Of course, the answer is just a straight line segment joining A and B.

10.3   THE EULER EQUATION

The historical problems from the previous section are particular instances of the general functional

$$I(y) = \int_a^b f(x, y, y')\, dx \tag{10.11}$$

where, in a typical **unconstrained** problem, the value of y is specified at $x = a$ and $x = b$, that is,

$$y(a) = y_a, \qquad y(b) = y_b. \tag{10.12}$$

If $I(y)$ has a maximum or minimum, its value may well depend on the properties of the space of admissible functions — enlarging this space tends to increase a maximum and decrease a minimum. Assume that y is a member of the set of continuously differentiable functions over the interval (a, b) satisfying $y(a) = y_a$, $y(b) = y_b$. Suppose that $I(y)$ has its maximum/minimum value when $y(x) = \eta(x)$ and let $h(x)$ be any continuously differentiable function over the interval (a, b) satisfying $h(a) = h(b) = 0$, then $y(x) = \eta(x) + \varepsilon h(x)$ satisfies the boundary conditions at $x = a$ and $x = b$. Furthermore, $I(\eta + \varepsilon h) = \bar{I}(\varepsilon)$ is a function of ε with a maximum/minimum turning value at $\varepsilon = 0$. Thus $\bar{I}'(0) = 0$ (the prime denotes differentiation with respect to ε) for *any* suitable function h.

In particular, if $\bar{I}''(0) > 0$ then $y(x) = \eta(x)$ gives normally a minimum value for $I(y)$ whereas if $\bar{I}''(0) < 0$ leads normally to a maximum value for the functional when $y(x) = \eta(x)$. One can not be too strict about these claims because of the simplistic nature of the assumed functional variation about $y(x) = \eta(x)$. The stated condition is necessary but not sufficient to ensure that the extremum is a minimum or maximum, unlike its counterpart for a single variable.

Hence

$$I(\eta + \varepsilon h) = \bar{I}(\varepsilon) = \int_a^b f(x, \eta + \varepsilon h, \eta' + \varepsilon h')\, dx, \qquad (10.13)$$

$$\frac{d\bar{I}}{d\varepsilon} = \int_a^b (f_y h + f_{y'} h')\, dx, \qquad (10.14)$$

$$\frac{d^2\bar{I}}{d\varepsilon^2} = \int_a^b \left(f_{yy} h^2 + 2 f_{yy'} h h' + f_{y'y'} h'^2 \right)\, dx, \qquad (10.15)$$

where all partial derivatives are evaluated at $y = \eta + \varepsilon h$, and $y' = \eta' + \varepsilon h'$. In particular,

$$\begin{aligned}
\bar{I}'(0) &= \int_a^b (f_\eta h + f_{\eta'} h')\, dx = \int_a^b f_\eta h\, dx + \left[f_{\eta'} h \right]_a^b - \int_a^b \left[\frac{d}{dx}(f_{\eta'}) \right] h\, dx \\
&= \int_a^b \left[f_\eta - \frac{d}{dx}(f_{\eta'}) \right] h\, dx = \int_a^b \alpha(x) h(x)\, dx \qquad (10.16)
\end{aligned}$$

since $h(a) = h(b) = 0$ and where $\alpha(x)$ was just introduced as a short–hand notation. The reader may recall that $\bar{I}'(0) = 0$ is a requirement for stationary values of $I(y)$. Suppose $\alpha(x) \neq 0$, say positive, at some point of the interval $[a, b]$. There exists an interval $[c, d] \subset [a, b]$ over which $\alpha(x) > 0$. Take

$$h(x) = \begin{cases} (x - c)(d - x) & x \in [c, d] \\ 0 & x \in [a, b] - [c, d] \end{cases}$$

then

$$\int_a^b \alpha(x) h(x)\, dx = \int_c^d \alpha(x)(x - c)(d - x)\, dx > 0$$

which is a contradiction. Hence a necessary condition for $I(y)$ to have a maximum/minimum value is that $\eta(x)$ satisfies the differential equation

$$\frac{\partial f}{\partial y} - \frac{d}{dx}\left(\frac{\partial f}{\partial y'}\right) = 0 \qquad (10.17)$$

and the boundary conditions $\eta(a) = y_a$, $\eta(b) = y_b$. Equation (10.17) is called the **Euler equation**[1] for the functional $I(y)$. A similar argument applies to a minimum. A solution of the Euler equation is called an **extremal** and the value of the functional on an extremal curve [often denoted by $y_0(x)$] is called an **extremum**. In the vocabulary of functions of a single variable, an *extremal* is the equivalent of a stationary point and an *extremum* is the equivalent of the associated stationary value. For the type of functional defined in (10.11), the Euler equation is a differential equation of second order.

Remarks 10.1

1. Suppose that $y = y(x)$ has continuous first derivative and satisfies the Euler equation (10.17). If $f(x, y, y')$ has continuous first and second order derivatives with respect to all its arguments, then $y(x)$ has a continuous second derivative wherever $f_{y'y'} \neq 0$. We do not prove this result but it serves notice that continuously differentiable extremals may not have continuous curvature.

2. A similar argument can be applied to the general functional

$$I(y) = \int_a^b f(x, y, y', y'', \ldots, y^{(n)})\, dx$$

where $y(x)$ satisfies the boundary conditions

$$y(a) = A_0, \quad y'(a) = A_1, \quad y''(a) = A_2, \quad \ldots, \quad y^{(n-1)}(a) = A_{n-1},$$
$$y(b) = B_0, \quad y'(b) = B_1, \quad y''(b) = B_2, \quad \ldots, \quad y^{(n-1)}(b) = B_{n-1}.$$

Here the extremals of $I(y)$ are solutions of the Euler equation

$$\frac{\partial f}{\partial y} - \frac{d}{dx}\left(\frac{\partial f}{\partial y'}\right) + \frac{d^2}{dx^2}\left(\frac{\partial f}{\partial y''}\right) + \ldots + (-1)^n \frac{d^n}{dx^n}\left(\frac{\partial f}{\partial y^{(n)}}\right) = 0$$

which is a differential equation of order $2n$. △

---

[1] It will be consistently referred to as *the* Euler equation to avoid any confusion with *Euler's differential equation* or *differential equations of Euler's type* treated in Section 4.6.

10.4   SUFFICIENT CONDITIONS

For the vast majority of applications, the nature of the extremum is controlled by the properties of

$$\bar{I}''(0) = \int_a^b (f_{\eta\eta} h^2 + 2 f_{\eta\eta'} h h' + f_{\eta'\eta'} h'^2)\, dx \qquad (10.18)$$

which follows from (10.15) by setting $\varepsilon = 0$. We let h be any non–zero continuously differentiable function satisfying $h(a) = h(b) = 0$. Whenever $\bar{I}''(0) > 0$ for all admissible h, then the extremum is a **minimum** and whenever $\bar{I}''(0) < 0$ for all admissible h, then the extremum is a **maximum**, otherwise it is indeterminate.

(a) If the integrand of (10.18) is a positive/negative definite quadratic form in h and h' then a conclusion can be drawn immediately.

(b) Assuming suitable differentiability conditions on η and all the functions appearing in the integral expression for $\bar{I}''(0)$, we have

$$\int_a^b 2 f_{\eta\eta'} h h'\, dx \;=\; [f_{\eta\eta'} h^2]_a^b - \int_a^b \left[\frac{d}{dx}(f_{\eta\eta'}) \right] h^2\, dx$$

$$= \; - \int_a^b \left[\frac{d}{dx}(f_{\eta\eta'}) \right] h^2\, dx$$

and so

$$\bar{I}''(0) = \int_a^b \left[\left(f_{\eta\eta} - \frac{d}{dx} f_{\eta\eta'} \right) h^2 + f_{\eta'\eta'} h'^2 \right] dx . \qquad (10.19)$$

As previously, a conclusion can be drawn depending on the nature of the coefficients of this integrand. Formula (10.19) provides another version of (10.18).

10.5   SPECIALIZATIONS OF THE EULER EQUATION

Dependent on the form of $f(x, y, y')$, three special cases exist in which the problem of finding extrema is significantly simplified.

Case 1 — missing y. Suppose that f does not contain y explicitly. In this event the functional is

$$I(y) = \int_a^b f(x, y')\, dx \qquad (10.20)$$

and the Euler equation is

$$\frac{d}{dx}\left(\frac{\partial f}{\partial y'} \right) = 0 \qquad \longrightarrow \qquad \frac{\partial f}{\partial y'} = C \qquad (10.21)$$

where C is a constant. Hence the Euler equation is reduced to an *algebraic equation* for y' in terms of x which can be solved by *quadrature*.

Case 2 — missing x. Suppose that f does not contain x explicitly. In this event the functional is

$$I(y) = \int_a^b f(y, y') \, dx.$$

(10.22)

Functionals of this type occur often in physical problems. If f satisfies the Euler equation (10.17) then

$$\frac{d}{dx}(f - y' f_{y'}) = f_y y' + f_{y'} y'' - y'' f_{y'} - y' \frac{d}{dx}(f_{y'})$$

$$= y' \left[f_y - \frac{d}{dx}(f_{y'}) \right] = 0.$$

(10.23)

Hence the Euler equation has **first integral**

$$f - y' f_{y'} = C$$

(10.24)

where C is a constant. This is now an *algebraic* equation for y' in terms of y and once again quadrature leads to the solution.

Case 3 — missing y'. Suppose that f does not contain y' explicitly. In this event the functional is

$$I(y) = \int_a^b f(x, y) \, dx$$

(10.25)

and the Euler equation is $f_y = 0$ which is an *implicit relationship* for y in terms of x. Here no arbitrary constants arise since no integration is performed and so it remains to check that the boundary conditions at $x = a$ and $x = b$ are satisfied.

10.6   SELECTED VARIATIONAL PROBLEMS

Example 10.1   Find the extremal and the extremum of the functional

$$I(y) = \int_1^8 \left[9 \left(\frac{dy}{dx} \right)^2 - \frac{2y^2}{x^2} \right] dx$$

for which $y(1) = 1$ and $y(8) = 4$.

Solution 10.1   Clearly $f_y = -4y/x^2$ and $f_{y'} = 18y'$ leading to the Euler equation

$$9x^2 \frac{d^2 y}{dx^2} + 2y = 0.$$

This is a differential equation of Euler's type (see Section 4.6) and has solutions of the form x^λ where

$$9\lambda(\lambda-1)+2=0 \quad\longrightarrow\quad (\lambda-1/3)(\lambda-2/3)=0 \quad\longrightarrow\quad \lambda_1=1/3,\ \lambda_2=2/3.$$

Therefore the Euler equation has general solution

$$y(x) = Ax^{1/3} + Bx^{2/3}$$

where A and B are constants to be fixed by the requirements $y(1)=1$ and $y(8)=4$. Hence

$$\begin{aligned}
y(1) &= 1 \ \longrightarrow\ & A+B=1 \\
y(8) &= 4 \ \longrightarrow\ & 2A+4B=4
\end{aligned} \quad\longrightarrow\ A=0,\ B=1.$$

Thus, the extremal curve is $y_0(x) = x^{2/3}$. The extremum is calculated to be

$$I\left(x^{2/3}\right) = \int_1^8 \left[9\left(\frac{2}{3}x^{-1/3}\right)^2 - 2\frac{x^{4/3}}{x^2} \right] dx$$

$$= \int_1^8 2x^{-2/3}\,dx = 2\left[3x^{1/3}\right]_1^8 = 6. \qquad\square$$

Example 10.2   Solve a *Brachistochrone* problem assuming that the particle starts from rest at $(-a, y_a)$ and falls to the origin $(a>0,\ y_a>0)$.

Solution 10.2   The time taken to traverse a path from $(-a, y_a)$ to $(0,0)$ starting from rest appears as a specialization of (10.5) upon setting $x_A=0$, $x_B=-a$, $y_A=0$, $y_B=y_a$, $v_A=0$. The functional to be minimized is

$$\tau(y) = \int_{-a}^0 \sqrt{\frac{1+y'^2}{2g(y_a-y)}}\,dx\,, \qquad y(-a)=y_a\,, \quad y(0)=0.$$

This functional is covered by Case 2 in Section 10.5 where x is missing from f and so the Euler equation has first integral, see (10.24),

$$\left(\sqrt{1+y'^2} - \frac{y'y'}{\sqrt{1+y'^2}}\right)\frac{1}{\sqrt{y_a-y}} = C \quad\longrightarrow\quad (y_a-y)(1+y'^2)=A$$

where A is a constant. Let $w = y_a - y$ then the previous equation can be reformulated as

$$w(1+w'^2) = A \qquad\longrightarrow\qquad w' = \frac{\sqrt{A-w}}{\sqrt{w}}\,, \quad w(-a)=0,\ w(0)=y_a\,.$$

Now let $w = A \sin^2 \theta$ for a new dependent variable θ so that

$$w' = \frac{dw}{d\theta}\frac{d\theta}{dx} = \sqrt{\frac{A - A\sin^2\theta}{A\sin^2\theta}} = \cot\theta \qquad \longrightarrow \qquad \frac{dx}{d\theta} = 2A\sin^2\theta.$$

This equation can be integrated to find $x(\theta)$ and leads to the solution

$$x = -a + A(\theta - \sin\theta\cos\theta), \qquad y = y_a - A\sin^2\theta$$

where the arbitrary constant A is to be found from the condition that this parametric curve passes through the origin. Note that when $\theta = 0$, the specified curve automatically passes through the starting point $(-a, y_a)$. Let the origin correspond to $\theta = \alpha \ (> 0)$, then A and α are determined from the relations

$$-a + A(\alpha - \sin\alpha\cos\alpha) = 0, \qquad y_a = A\sin^2\alpha.$$

The constant A can be eliminated between these equations so that α satisfies the transcendental equation

$$y_a(\alpha - \sin\alpha\cos\alpha) - a\sin^2\alpha = 0.$$

Further, the extremal time is

$$\tau = \int_{-a}^{0} \sqrt{\frac{1 + y'^2}{2g(y_a - y)}}\, dx = \int_{0}^{\alpha} \sqrt{\frac{x_\theta^2 + y_\theta^2}{2g(y_a - y)}}\, d\theta$$

with the short–hand notation $x_\theta = dx/d\theta$ and $y_\theta = dy/d\theta$. Now

$$x_\theta^2 + y_\theta^2 = 4A^2\sin^4\theta + 4A^2\sin^2\theta\cos^2\theta = 4A^2\sin^2\theta,$$

and so the extremal time becomes

$$\tau = \int_{0}^{\alpha} \sqrt{\frac{4A^2\sin^2\theta}{2gA\sin^2\theta}}\, d\theta = \sqrt{\frac{2A}{g}}\,\alpha = \sqrt{\frac{2y_a}{g}}\,\frac{\alpha}{\sin\alpha}. \qquad \square$$

Remarks 10.2   The solution to Example 10.2 is part of an inverted **cycloid**. Consider a wheel of radius a rolling along a horizontal surface. Mark a point on the rim of this wheel and take θ to be the angle between the vertical direction and the mark, subtended at the centre of the wheel. The marked point traces out a cycloid as the wheel rolls along. $\triangle$

Example 10.3   Find the extremal of the functional

$$I(y) = \int_{-1}^{1} y^2(2x - y')^2\, dx, \qquad y(-1) = 0, \quad y(1) = 1.$$

Solution 10.3   Here $f_{y'y'} = -2y^2$ and this could be zero and so there is no reason to believe that in this case y'' should be continuous in the interval $[-1, 1]$. Indeed it can be seen that the extremum value is zero and is attained with extremal

$$y(x) = \begin{cases} 0 & -1 \le x \le 0 \\ x^2 & 0 \le x \le 1. \end{cases}$$

Of course, the Euler equation still applies and is satisfied.   □

Example 10.4   Of all the curves which join two fixed points with given coordinates (x_0, y_0) and (x_1, y_1), find the one which generates the surface of minimum area when revolved about the x axis.

Solution 10.4   The area of the surface of revolution is

$$I(y) = 2\pi \int_{x_0}^{x_1} y\sqrt{1 + y'^2}\, dx$$

which has first integral

$$f - y' f_{y'} = y\sqrt{1 + y'^2} - \frac{y'y'y}{\sqrt{1 + y'^2}} = A \qquad \longrightarrow \qquad \frac{y}{\sqrt{1 + y'^2}} = A$$

$$\rightarrow y'^2 = \frac{y^2}{A^2} - 1 \quad \longrightarrow \quad \int dx = \int \frac{A\, dy}{\sqrt{y^2 - A^2}} \quad \longrightarrow \quad x + a = A\cosh^{-1}\frac{y}{A}.$$

Thus the extremal curves are arcs of the *catenary*

$$y(x) = A\cosh\left(\frac{x + a}{A}\right)$$

where the integration constants a and A are determined by the conditions $y(x_0) = y_0$, $y(x_1) = y_1$. Of course, given any pair of points, there is no guarantee that it is possible to find real values for a and A such that the catenary passes through the given points. Indeed it is clear from $f_{y'y'}$ that not all extremals need have continuous second derivatives. There are three cases to consider.

(a) If a single catenary can be drawn through the two fixed points then it represents the solution to the problem.

(b) If two extremals can be drawn through the given points then *one* of these is the solution to the problem.

(c) If no single catenary passes through both points then the extremal is formed from the union of catenaries and straight line segments. Calculus of variations provides criteria (so–called Weierstrass–Erdmann corner conditions) which determine how the component curves are connected to form a full solution.   □

Example 10.5   Prove that

$$I(y) = \int_0^1 (y'^2 + y^2 + 2ye^{2x})\, dx \geq \frac{11}{36} (e^4 - 1)$$

where the extremum is calculated over the class of continuously differentiable functions satisfying $y(0) = 1/3$, $y(1) = e^2/3$.

Solution 10.5   The Euler equation of this functional is

$$y'' - y = e^{2x} \qquad \longrightarrow \qquad y(x) = Ae^x + Be^{-x} + \frac{e^{2x}}{3}.$$

The boundary conditions are satisfied with $A = B = 0$ and so the extremal curve is $y_0(x) = e^{2x}/3$. The extremum value is

$$I(y_0) = \int_0^1 \left(\frac{4e^{4x}}{9} + \frac{e^{4x}}{9} + \frac{2e^{4x}}{3} \right) dx = \frac{11}{9} \int_0^1 e^{4x}\, dx = \frac{11}{36} (e^4 - 1).$$

In order to establish that the extremum is a minimum we try a direct route instead of using one of the criteria (10.18) or (10.19) and consider

$$I(y_0 + h) - I(y_0) = \int_0^1 \left[(y_0' + h')^2 + (y_0 + h)^2 + 2e^{2x}(y_0 + h) \right.$$
$$\left. -y_0'^2 - y_0^2 - 2y_0 e^{2x} \right] dx$$
$$= \int_0^1 (h'^2 + h^2)\, dx + \frac{4}{3} \int_0^1 \left(e^{2x} h' + 2e^{2x} h \right) dx$$
$$= \int_0^1 (h'^2 + h^2)\, dx + \frac{4}{3} \left[e^{2x} h \right]_0^1 \qquad [h(0) = h(1) = 0]$$
$$= \int_0^1 (h'^2 + h^2)\, dx \geq 0$$

with equality if and only if $h \equiv 0$. This proves the original inequality.   □

10.7 CONSTRAINED EXTREMA

The introduction cited examples where one functional is minimized or maximized while another is constrained. For example, for the hanging chain of fixed length, the gravitational potential energy is minimized while the length remains fixed. Problems of that type lead to **constrained extrema**.

Consider the functional

$$I(y) = \int_a^b f(x, y, y') \, dx \qquad (10.26)$$

where the admissible curves $y(x)$ satisfy

$$y(a) = y_a, \qquad y(b) = y_b, \qquad J(y) = \int_a^b g(x, y, y') \, dx = K \qquad (10.27)$$

and $J(y)$ is another functional and K a constant. Let $I(y)$ have extremal curve $\eta(x)$ then there exists a constant λ, called a **Lagrange multiplier**, such that $y = \eta(x)$ is an extremal of the functional

$$I(y) = \int_a^b F(x, y, y') \, dx = \int_a^b \Big[f(x, y, y') + \lambda g(x, y, y') \Big] \, dx \qquad (10.28)$$

that is, $y = \eta(x)$ satisfies the *Euler equation*

$$\frac{\partial F}{\partial y} - \frac{d}{dx} \left(\frac{\partial F}{\partial y'} \right) = 0 \qquad (10.29)$$

and the conditions (10.27).

To provide some justification for this result, suppose that $y = \eta(x)$ is the extremum and consider $I(\eta + \varepsilon h + \alpha h_1)$ where ε and α are real numbers and h, h_1 are any two differentiable functions of x which are zero at $x = a$ and $x = b$. Moreover $J(\eta + \varepsilon h + \alpha h_1) = K$ and this condition effectively defines α as a function of ε, that is, $\alpha = \alpha(\varepsilon)$. Also, $\alpha(0) = 0$. Henceforth $I(\eta + \varepsilon h + \alpha h_1) = V(\varepsilon)$ and a necessary condition for $y = \eta(x)$ to be an extremal of $I(y)$ is that $V'(0) = 0$. Since

$$\int_a^b g\Big(x, \eta + \varepsilon h + \alpha(\varepsilon) h_1, \eta' + \varepsilon h' + \alpha(\varepsilon) h_1' \Big) \, dx = K,$$

differentiation with respect to ε yields

$$\int_a^b \left(g_y h + g_y h_1 \frac{d\alpha}{d\varepsilon} + g_{y'} h' + g_{y'} h_1' \frac{d\alpha}{d\varepsilon} \right) dx = 0$$

where the partial derivatives of g are evaluated at $y = \eta + \varepsilon h + \alpha(\varepsilon) h_1$ and $y' = \eta' + \varepsilon h' + \alpha(\varepsilon) h_1'$. In particular,

$$\frac{d\alpha}{d\varepsilon} \bigg|_{\varepsilon=0} = - \frac{\int_a^b (g_y h + g_{y'} h') \, dx}{\int_a^b (g_y h_1 + g_{y'} h_1') \, dx}$$

where all partial derivatives are now evaluated at $y = \eta(x)$ and $y' = \eta'(x)$. Hence

$$
\frac{d\bar{I}}{d\varepsilon} = \frac{d}{d\varepsilon} \int_a^b f\left(x, \eta + \varepsilon h + \alpha(\varepsilon)h_1, \eta' + \varepsilon h' + \alpha(\varepsilon)h_1'\right) dx
$$

$$
= \int_a^b \left(f_y h + f_y h_1 \frac{d\alpha}{d\varepsilon} + f_{y'} h' + f_{y'} h_1' \frac{d\alpha}{d\varepsilon}\right) dx
$$

$$
= \int_a^b \left(f_y h + f_{y'} h'\right) dx + \frac{d\alpha}{d\varepsilon} \int_a^b \left(f_y h_1 + f_{y'} h_1'\right) dx
$$

$$
= \int_a^b \left(f_y h + f_{y'} h'\right) dx - \frac{\displaystyle\int_a^b \left(g_y h + g_{y'} h'\right) dx}{\displaystyle\int_a^b \left(g_y h_1 + g_{y'} h_1'\right) dx} \int_a^b \left(f_y h_1 + f_{y'} h_1'\right) dx.
$$

Now choose *any* suitable function $h_1(x)$ satisfying $h_1(a) = h_1(b) = 0$, then from the previous expression for $\bar{I}'(\varepsilon)$, it follows immediately that

$$
\bar{I}'(0) = \int_a^b \left(f_\eta h + f_{\eta'} h'\right) dx + \lambda \int_a^b \left(g_\eta h + g_{\eta'} h'\right) dx
$$

where λ is a constant and denotes the ratio

$$
\lambda = -\frac{\displaystyle\int_a^b \left(f_\eta h_1 + f_{\eta'} h_1'\right) dx}{\displaystyle\int_a^b \left(g_\eta h_1 + g_{\eta'} h_1'\right) dx}.
$$

Since $\bar{I}'(0) = 0$ is a necessary condition for an extremal of I then it follows by the usual argument, see (10.16) in the unconstrained case, that $F = f + \lambda g$ is now a solution of the Euler equation (10.29).

Remarks 10.3

1. If only one constraint is imposed, there is only one *Lagrange multiplier* λ. More constraints can be incorporated into the problem provided a corresponding Lagrange multiplier is introduced for each extra constraint.

2. For one constraint there are three constants to be determined, namely λ and two constants of integration, and these are determined from the two boundary conditions and the integral constraint. Clearly each additional constraint introduces a further Lagrange multiplier which is in turn determined by the integral constraint. $\triangle$

Example 10.6   Consider the curve $y = \eta(x)$ of length $\pi/2$ which joins the points $A(0,2)$ and $B(1,3)$. Show that to give an extremum of the area enclosed by the curve, the line $x = 1$ and the x and y axes, the curve necessarily has constant curvature, say κ.

Without solving the Euler equation, show that

$$I(\eta + h) - I(\eta) = \frac{1}{2\kappa} \int_0^1 \frac{h'^2}{(1 + \eta'^2)^{3/2}} \, dx$$

for any function $h(x)$ satisfying $h(0) = h(1) = 0$. Hence find the extremum values for $I(\eta)$ and the corresponding extremal curves.

Solution 10.6   Here

$$I(y) = \int_0^1 y \, dx \,, \qquad J(y) = \int_0^1 \sqrt{1 + y'^2} \, dx = \frac{\pi}{2} \,.$$

Construct $F = y + \lambda\sqrt{1 + y'^2}$. The Euler equation (10.29) is

$$\frac{d}{dx}\left(\frac{\lambda y'}{\sqrt{1 + y'^2}}\right) = 1 \qquad \longrightarrow \qquad \frac{y''}{\left(1 + y'^2\right)^{3/2}} = \frac{1}{\lambda} \,.$$

By virtue of the standard formula for the curvature of a curve $y(x)$, we recognize that the required extremal has constant curvature $\kappa = 1/\lambda$ and it is therefore an arc of a circle. We have

$$I(\eta + h) - I(\eta) = \int_0^1 h(x) \, dx \,, \qquad h(0) = h(1) = 0$$

where $y(x) = \eta(x) + h(x)$ necessarily satisfies the length constraint

$$\int_0^1 \sqrt{1 + (\eta' + h')^2} \, dx = \frac{\pi}{2} \,.$$

The square root in the integral is now expanded in terms of h and h' to obtain

$$\sqrt{1 + (\eta' + h')^2} = \sqrt{1 + \eta'^2} + \frac{\eta' h'}{\sqrt{1 + \eta'^2}} + \frac{1}{2} \frac{h'}{(1 + \eta'^2)^{3/2}} + \cdots$$

where higher order terms in h and h' have been ignored. Therefore the constraint condition requires that

$$\int_0^1 \sqrt{1 + \eta'^2} \, dx + \int_0^1 \frac{\eta' h'}{\sqrt{1 + \eta'^2}} \, dx + \frac{1}{2} \int_0^1 \frac{h'^2}{(1 + \eta'^2)^{3/2}} \, dx + \cdots = \frac{\pi}{2} \,.$$

Since the curve $\eta(x)$ automatically satisfies the constraint condition, integration by parts yields

$$\int_0^1 \frac{\eta' h'}{\sqrt{1+\eta'^2}}\, dx = \left[\frac{\eta' h}{\sqrt{1+\eta'^2}}\right]_0^1 - \int_0^1 \left(\frac{\eta'}{\sqrt{1+\eta'^2}}\right)' h\, dx = -\kappa \int_0^1 h\, dx .$$

In conclusion,

$$I(\eta + h) - I(\eta) = \frac{1}{2\kappa} \int_0^1 \frac{h'^2}{(1+\eta'^2)^{3/2}}\, dx$$

(exact to second order). There are now two possible ways to draw a circle passing through the points A and B.

(a) If $\kappa > 0$ then $I(\eta + h) > I(\eta)$ and the extremal

$$\eta(x) = 3 - \sqrt{1-x^2}$$

gives a minimum extremum value of $(3 - \pi/4)$.

(b) If $\kappa < 0$ then $I(\eta + h) < I(\eta)$ and the extremal

$$\eta(x) = 2 + \sqrt{x(2-x)}$$

gives a maximum extremum value of $(2 + \pi/4)$.   □

Example 10.7   Find the extremal curve joining the two points $(0,0)$ and $(1,0)$, enclosing unit area with the x axis and which yields an extremum of the functional

$$I(y) = \int_0^1 y'^2\, dx .$$

Calculate the extremum value and verify that it is a minimum.

Solution 10.7   The constraint for this problem is

$$J(y) = \int_0^1 y\, dx = 1 .$$

Hence $F = y'^2 + \lambda y$ and the Euler equation has first integral $F - y'F_{y'} = A$ leading to

$$y'^2 + \lambda y - 2y'^2 = A \longrightarrow y'^2 = \lambda y - A \longrightarrow \int \frac{dy}{\sqrt{\lambda y - A}} = x + B .$$

After integration, the extremal curves are seen to be

$$\lambda^2(x + B)^2 = 4(\lambda y - A) \qquad \longrightarrow \qquad 4\lambda y = \lambda^2(x + B)^2 + 4A \,.$$

We have

$$y(0) = 0 \qquad \longrightarrow \qquad \lambda^2 B^2 + 4A = 0 \qquad \longrightarrow \qquad 4y = \lambda(x^2 + 2Bx) \,,$$

$$y(1) = 0 \qquad \longrightarrow \qquad 2B + 1 = 0 \qquad \longrightarrow \qquad y = (\lambda/4)x(x-1)$$

where λ is determined by the condition

$$1 = \int_0^1 y \, dx = \frac{\lambda}{4} \int_0^1 x(x-1) \, dx = -\frac{\lambda}{24} \,.$$

In conclusion, the extremal curve is the parabola $y_0(x) = 6x(1-x)$ and the extremum value is

$$I(y_0) = \int_0^1 {y_0'}^2 \, dx = 36 \int_0^1 (1 - 2x)^2 \, dx = 36 \left[\frac{(1-2x)^3}{-6} \right]_0^1 = 12 \,.$$

To see that this is a minimum, consider

$$I(y_0 + h) - I(y_0) = \int_0^1 \left[(y_0' + h')^2 - {y_0'}^2 \right] dx = \int_0^1 h'^2 \, dx + 2 \int_0^1 y_0' h' \, dx$$

where h is any function satisfying

$$h(0) = 0\,, \qquad h(1) = 0\,, \qquad \int_0^1 h(x) \, dx = 0$$

[the integral constraint for h follows from the fact that $J(y_0 + h) = 1$ and $J(y_0) = 1$]. On integration by parts,

$$\int_0^1 y_0' h' \, dx = \left[y_0' h \right]_0^1 - \int_0^1 h(x) y_0'' \, dx = 12 \int_0^1 h(x) \, dx = 0 \,.$$

In conclusion, $I(y_0 + h) \geq I(y_0)$ for all non–zero h, that is, $y = y_0(x)$ yields a minimum to $I(y)$. More often than not, it is a non–trivial matter to verify the nature of a extremum in a constrained problem.   □

Example 10.8   It was shown previously that the configuration of a chain of *uniform* linear density and fixed length, suspended between two fixed points (horizontally separated) minimizes the gravitational potential energy

$$V(y) = \rho g \int_{x_A}^{x_B} y \sqrt{1 + y'^2} \, dx \,.$$

In addition, y is constrained to satisfy

$$J(y) = \int_{x_A}^{x_B} \sqrt{1+y'^2}\, dx = l$$

where l is the length of the chain. Solve this problem for a chain that is suspended between the points $(-d, 0)$ and $(d, 0)$ where $l > 2d$.

Solution 10.8   Let

$$F = \rho g y \sqrt{1+y'^2} + \rho g \lambda \sqrt{1+y'^2} = \rho g (y+\lambda)\sqrt{1+y'^2}$$

(where we have introduced the Lagrange multiplier conveniently as $\rho g\lambda$), then the first integral $F - y'F_{y'} = C$ of the Euler equation is

$$(y+\lambda)\sqrt{1+y'^2} - y'\frac{(y+\lambda)y'}{\sqrt{1+y'^2}} = A \quad \longrightarrow \quad \frac{y+\lambda}{\sqrt{1+y'^2}} = A\,.$$

The method of solution closely mirrors a previous variational problem (the catenary problem of Example 10.4). The only difference here is that y is now replaced by $y + \lambda$. The appropriate solution is

$$x + a = A\cosh^{-1}\left(\frac{y+\lambda}{A}\right) \quad \longrightarrow \quad y+\lambda = A\cosh\left(\frac{x+a}{A}\right)$$

where constants a, A and λ are to be chosen so that

$$y(-d) = 0\,, \quad y(d) = 0\,, \quad J(y) = \int_{-d}^{d}\sqrt{1+y'^2}\,dx = l\,.$$

The two boundary conditions give

$$\lambda = A\cosh\left(\frac{a-d}{A}\right)\,, \quad \lambda = A\cosh\left(\frac{a+d}{A}\right)\,.$$

The last two relations require that $a = 0$ and so the extremal simplifies to $y = A\cosh(x/A) - A\cosh(d/A)$. Furthermore, the length of the chain is

$$l = \int_{-d}^{d}\cosh\left(\frac{x}{A}\right)dx = 2A\sinh\left(\frac{d}{A}\right)\,.$$

Let $\beta = l/2d > 1$ and define $z = d/A$. The constraint condition now states that z is the solution of the transcendental equation

$$\beta z = \sinh z\,, \quad \beta > 1\,.$$

Figure 10.4 indicates that this equation has exactly one solution if $\beta > 1$ (and the trivial solution only if $0 < \beta < 1$). Let the root be denoted by z_0, then we have $A = d/z_0$ and therefore we obtain the extremal curve as

$$y_0(x) = \frac{d}{z_0}\left[\cosh\left(\frac{xz_0}{d}\right) - \cosh z_0\right]\,. \qquad \square$$

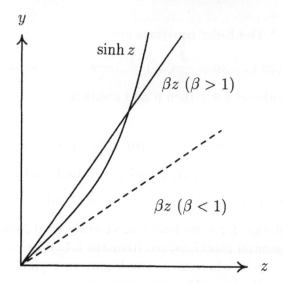

Figure 10.4: Graphical solution of $\beta z = \sinh z$. Shown are cases for $\beta < 1$ (dashed line) and $\beta > 1$ (full line).

10.8 FUNCTIONALS OF SEVERAL VARIABLES

Functionals involving many dependent functions $y_1, y_2, \ldots, y_n$ but only one independent variable, say t, occur often in practice. Here only the extremal problem

$$I(y_1, y_2, \ldots, y_n) = \int_a^b f(t; y_1, y_2, \ldots, y_n; \dot{y}_1, \dot{y}_2, \ldots, \dot{y}_n)\, dt \qquad (10.30)$$

is considered. For example, the equations of classical mechanics are obtained from a functional of this form called the **Lagrangian**. The derivation of the Euler equations for this functional mirrors exactly the derivation of the standard Euler equation but applied to each independent component function of I. A necessary condition for an extremal of the functional $I(y_1, \ldots, y_n)$ with a and b fixed and $y_k(a) = A_k$, $y_k(b) = B_k$ $(k = 1, 2, \ldots, n)$ is that the functions $y_k = y_k(t)$, $(k = 1, 2, \ldots, n)$ satisfy the *Euler equations*

$$\frac{\partial f}{\partial y_k} - \frac{d}{dt}\left(\frac{\partial f}{\partial \dot{y}_k}\right) = 0, \qquad k = 1, 2, \ldots, n. \qquad (10.31)$$

Example 10.9   Find curves $y = y(x)$, $z = z(x)$ for which the functional

$$I(y, z) = \int_0^{\pi/2} (y'^2 + z'^2 + 2yz)\, dx$$

has an extremum, given that $y(0) = z(0) = 0$ and $y(\pi/2) = z(\pi/2) = 1$.

Solution 10.9   The Euler equations are

$$2z = \frac{d}{dx}(2y'), \quad 2y = \frac{d}{dx}(2z') \qquad \longrightarrow \qquad y'' = z, \quad z'' = y.$$

Let $u = y - z$ and $v = y + z$ then u and v satisfy

$$u'' + u = 0, \qquad u(0) = 0, \quad u(\pi/2) = 0,$$
$$v'' - v = 0, \qquad v(0) = 0, \quad v(\pi/2) = 2.$$

Hence arbitrary constants A, B, C and D can be found so that

$$u(x) = A \sin x + B \cos x, \qquad v(x) = C \sinh x + D \cosh x$$

where in the solution for v we have conveniently used hyperbolic functions instead of exponential functions, see Remarks 3.1. The conditions on u are such that $A = B = 0$ whereas the conditions on v yield $C = 2/\sinh(\pi/2)$ and $D = 0$. The solution is therefore given by

$$y(x) = z(x) = \frac{\sinh x}{\sinh(\pi/2)} \, . \qquad\qquad \square$$

10.9   TUTORIAL EXAMPLES 10

T 10.1   Find the extremal of

$$I(y) = \int_0^3 (y'^2 + 4xy)\, dx$$

when $y(0) = 0$, $y(3) = 12$ and verify that this extremal minimizes I.

T 10.2   Find the extremal of

$$I(y) = \int_0^1 \left(y^2 + y'^2 - 4xy \right) dx$$

for which $y(0) = 0$ and $y(1) = 2$. Evaluate the extremum and show that this is a minimum by confirming that $\bar{I}''(0) \geq 0$ where

$$\bar{I}''(0) = \int_0^1 \left(f_{yy}\, h^2 + 2f_{yy'}\, hh' + f_{y'y'}\, h'^2 \right) dx$$

for any continuously differentiable function $h(x)$.

T 10.3   Show that the Euler equation of the functional

$$I(y) = \int_a^b f(x, y, y')\, dx$$

is satisfied identically provided there exists a suitably differentiable function $\phi(x, y)$ such that $f(x, y, y') = \phi_x + \phi_y y'$. Explain the meaning of this result so far as the functional is concerned.

T 10.4   Find the extremal $y_0(x)$ and extremum of the functional

$$\int_0^{\pi/4} (y^2 - y'^2 + 6y \sin 2x)\, dx\,, \qquad y(0) = 0, \quad y(\pi/4) = 1\,.$$

A function $h(x)$ is defined on the interval $[0, a]$ by

$$h(x) = \sin \lambda x \int_x^a \frac{f(u)}{\sin \lambda u}\, du$$

where λ, a are constants such that $0 < \lambda < \pi/a$ and $f(t)$ is defined in terms of the continuous function ϕ by

$$f(t) = -\frac{1}{\sin \lambda t} \int_0^t \phi(u) \sin \lambda u\, du\,.$$

Show that h satisfies the differential equation $h'' + \lambda^2 h = \phi$ and the boundary conditions $h(0) = h(a) = 0$. Verify that

$$\int_0^a h(h'' + \lambda^2 h)\, dx = -\int_0^a f^2(t)\, dt$$

and deduce further that

$$\int_0^a h'^2\, dx \geq \lambda^2 \int_0^a h^2\, dx\,.$$

Hence prove that for all continuously differentiable functions $y(x)$ on the interval $(0, \pi/4)$ satisfying $y(0) = 0$, $y(\pi/4) = 1$, the extremal $y_0(x)$ gives a maximum of the functional $I(y)$.

T 10.5   Determine the extremal $y_0 = y_0(x)$ of the functional

$$I(y) = \int_0^1 \left(\frac{1 + y^2}{y'^2} \right) dx\,, \qquad y(0) = 0, \quad y(1) = a\,.$$

Show that $|a| < 1$ is a sufficient condition for a minimum.

Appendix A

Self–study Projects

The purpose of including the following projects in this book is to provide an opportunity for extended individual or small group work that goes beyond the examples embedded into the text and the tutorial examples of the ten chapters. The six projects outlined below have been used in recent years in third year mathematics at the Department of Mathematics, University of Glasgow, whereby groups of three to four students worked together towards the solution of an individual project. On completion, usually after a period of some weeks, an oral presentation of the project's aims, tasks and results was delivered. While there is no reason that individual students or small groups of students may not achieve the aims of the project completely by their own efforts, it is more common that some guidance from a supervisor will be sought. For that reason there is no necessity here to include the project solutions.

A.1   ROCKET POWER

Project summary: This project examines the flight of a rocket that ascends and descends vertically under its own power.

Introduction

Newton's Second Law states that the rate of change of the momentum of a body is equal to the applied forces. Therefore

$$\frac{dp}{dt} = F_{\text{ext}} \tag{A.1}$$

where F_{ext} summarizes the applied forces and the momentum p is given by

$$p = Mv, \tag{A.2}$$

M being the mass of the object and v its velocity. In many problems the mass is constant and (A.1) simplifies to

$$M\frac{dv}{dt} = F_{\text{ext}} \qquad (A.3)$$

which can then be solved for the velocity v when the forces are given.

In this project we shall treat the more general case where the mass M is not a constant but varies with time, i.e., the simplification from (A.1) to (A.3) does not apply. The particular problem to be investigated is that of a rocket which ascends vertically under its own power.

Let $v(t)$ be the velocity and $M(t)$ the mass of the rocket at time t. The rocket's momentum is therefore $p(t) = M(t)v(t)$. As time passes, it burns propellant and ejects it at the exhaust with a constant velocity u (relative to the rocket). After a short time δt the rocket has ejected δm mass of propellant. Consequently, the rocket's mass is $M(t + \delta t) = M(t) - \delta m$ and its velocity has changed to $v(t+\delta t) = v(t)+\delta v$. In order to avoid calculating the force on the rocket due to fuel burning it is easier to consider the physical system (our 'body') to consist of the rocket and its fuel. In the first instance, show that the change in momentum of that body between t and $t + \delta t$ is given by

$$p(t + \delta t) - p(t) = M\delta v - \delta m\,\delta v - u\,\delta m. \qquad (A.4)$$

Consequently, determine the rate of change of the momentum as $\delta t \to 0$. The arising expression must equal the sum of the external forces. Assume here that two external forces are acting on the rocket, namely (i) **gravity**, which is assumed to act downward and (ii) **air resistance**, assumed proportional to the velocity of the rocket but opposing the direction of motion.

Show that the differential equation governing the motion of the rocket is given by

$$M(t)\frac{dv(t)}{dt} - u\frac{dm(t)}{dt} = -g\,M(t) - k\,v(t) \qquad (A.5)$$

where $g = 9.81$ m/s^2 is the gravitational acceleration and k is a proportionality constant. Discuss (A.5) with particular emphasis as to whether it provides a self–consistent equation to model the flight of the rocket.

In addition to (A.5) we need to include the relation

$$\frac{dm(t)}{dt} = -\frac{dM(t)}{dt}. \qquad (A.6)$$

What is its significance?

Burnout and beyond

Assume that fuel is consumed at a constant rate

$$\frac{dm(t)}{dt} = f .$$

(A.7)

Solve (A.6) for $M(t)$ if the total mass (rocket including fuel) at $t = 0$ is M_0. Subsequently, solve (A.5) for a rocket which is taking off at $t = 0$.

Consider now some data pertaining to a rocket: initial weight $M_0 = 12800$ kg, fuel weight 8800 kg, fuel consumption rate $f = 125$ kg/s, exhaust velocity $u = 2000$ m/s, air resistance parameter $k = 1.5$ kg/s and thus address the following tasks:

1. Find the time of **burnout**, i.e., when all fuel reserves are expended.

2. Calculate the **burnout velocity** defined as the velocity at burnout.

3. What height has the rocket reached at burnout?

4. After burnout, with all its fuel spent, the rocket continues to rise subject to the force of gravity and the air resistance. Find the time at which it reaches the highest point on its trajectory.

5. What is the rocket's maximum distance above the surface of the earth?

6. From its highest elevation above ground, the rocket falls to the ground. Find the approximate time of impact on the surface and the corresponding impact velocity.

7. Draw diagrams of the velocity and the acceleration (that is the rate of change of the velocity) as functions of time.

8. Discuss the limits of the above model particularly with respect to the way gravity and air resistance have been modelled in (A.5). Say, a satellite is to be put into an orbit at about 300 km above the Earth's surface. Decide from **Newton's Law of gravitation** whether it is justified to assume, as has been done in (A.5), that the force of gravity is constant with height. Similarly, can you say something about the modelling of the air resistance term?

A burning program

A **burning program** is a function $f_{bp}(t)$ specifying the propellant consumption rate

$$f_{bp}(t) = \frac{dm(t)}{dt} .$$

(A.8)

Its integral is the mass of the fuel which has been expelled from the rocket (which is equal to the mass lost by the rocket). From (A.6) we have

$$m(t) + M(t) = M_0 \tag{A.9}$$

if the mass of rocket and fuel is M_0 at $t = t_0$. Now, look at (A.5) as a differential equation for $m(t)$. The objective is to find a burning program $f_{bp}(t)$ in such a way, that the burning program — when initiated at some time t_0 — gives a constant velocity $v(t) = v(t_0)$ to the rocket for all $t \geq t_0$.

A.2 RESONANCE EFFECTS IN BEAMS

Project summary: A mathematical model of a bridge is constructed in order to determine its natural oscillation frequency.

Introduction

The response of a structure to forced vibrations is now an important consideration in their engineering design. Most people are familiar with the story about soldiers breaking cadence when they march over a bridge but are perhaps less aware that the structure in question was the Broughton suspension bridge near Manchester, Great Britain, and that the event took place as long ago as 1831. The failure occurred when a column of soldiers marched over the bridge, thereby establishing a periodic force of large magnitude along its entire length. Unfortunately, the frequency of this force was effectively equal to the natural frequency of the bridge and, in the absence of a damping mechanism, very large oscillations were induced leading to collapse.

The collapse of the Tacoma Bridge across Puget Sound (in the State of Washington, USA) on 7 November 1940, is perhaps the most notable structural failure in modern times and was caused by vortex shedding alternating from above and below the bridge platform. Although this phenomenon is due to wind action and is unavoidable, in this instance, the shedding occurred at a frequency close to that of the structure's natural frequency.

This project aims to model a bridge platform as a one–dimensional beam. You will be required to formulate the differential equation which controls the oscillations and deduce the natural frequencies of particular configurations. Finally, you should try to obtain real data on a typical bridge and attempt to determine its natural frequency using, for example, *Maple*, *Mathematica*, or some other mathematical tools.

Suspension bridge

Have you ever wondered what shape is adopted by the large steel cables passing over the towers of a suspension bridge and supporting the platform below? Suppose that the platform has linear density ρ per unit length and is supported by a continuum of fine strands which run vertically and connect the steel cables and platform below. Formulate a differential equation for the shape of the cables, solve this equation and find the required shape.

Of course, the platform interacts with the steel cables via the vertical strands and therefore the mechanical properties of the platform influence the motion of the cables and the relative inextensibility of the cables likewise affects the motion of the platform. The real problem is complex and so as a first model, consider the situation in which the deformation of the platform is essentially isolated from that of the supporting structure. In this event, the platform can be modelled as a beam in the way outlined below.

Model equations

Consider a uniform one–dimensional beam of length l, composed of a material with *Young's Modulus* E, cross–sectional *moment of inertia* I and with density ρ per unit length. Suppose that $u(x,t)$ is the displacement of the beam from its horizontal equilibrium position and that it is subject to a driving force $f(x,t)$, then it can be shown that the motion of the beam minimizes the functional

$$L(u) = \int \int \left[\frac{\rho}{2} \left(\frac{\partial u}{\partial t} \right)^2 + f(x,t)u - \frac{EI}{2} \left(\frac{\partial^2 u}{\partial x^2} \right)^2 \right] \, dx \, dt \,. \qquad (A.10)$$

Can you explain the meaning of the terms in this functional? Deduce that the deformation of the beam in the presence of an applied load is governed by the partial differential equation

$$\rho \frac{\partial^2 u}{\partial t^2} + EI \frac{\partial^4 u}{\partial x^4} = f(x,t), \qquad 0 < x < l, \quad t > 0. \qquad (A.11)$$

Suppose that the beam is subject to an external periodic force given by $f(x,t) = f(x) \cos pt$. Time–periodic solutions of (A.11) are sought in the form $u(x,t) = y(x) \cos pt$. Introduce the new variable $z = x/l$ and deduce that $y(z)$ satisfies the fourth order ordinary differential equation

$$\frac{d^4 y}{dz^4} - \omega^2 y = g(z), \qquad \omega = pl^2 \sqrt{\rho/EI} \qquad (A.12)$$

in which $g(z)$ is a multiple of $f(x)$. The solution is completed by supplying appropriate boundary conditions at $z = 0$ and $z = 1$. The three most common boundary conditions are described in Table A.1 and can be applied in any combination, one pair at each end. Hence it is required to find non–zero solutions of (A.12) satisfying four boundary conditions, two at each end.

type of support	boundary conditions
freely supported	$y = y'' = 0$
cantilever supported	$y = y' = 0$
unsupported	$y'' = y''' = 0$

Table A.1: Common boundary conditions for vibrating beams.

When $g(z) \equiv 0$ then (A.12) is an eigenvalue problem. Rather interestingly, many of these problems have analytical solutions which in turn require $\lambda = \sqrt{\omega}$ to be the solution of some transcendental equation, for example, $\tanh \lambda = \tan \lambda$ or $\cosh \lambda \cos \lambda + 1 = 0$. Once a transcendental equation is isolated, *Maple* or *Mathematica* (or an equivalent tool) can be employed to estimate the first few solutions numerically. These values are then converted back into physical space and interpreted suitably.

Solve this eigenvalue problem for free–free, cantilever–unsupported, cantilever–free and cantilever–cantilever beams by starting with the general solution based on the solution of the auxiliary equation. Hence determine the lowest natural frequency of oscillation in each case.

A.3 SNOWPLOUGHS

Project summary: The village of Applecross is situated in a remote area of the west coast of Scotland. During the winter the single–track road to Kinlochewe (40 miles away) is always heavily affected by snow. In this project we shall consider some mathematical models of snow clearance with snowploughs.

Scenario 1

On a certain day it begins to snow early in the morning (throughout the whole region), and it continues to snow at a constant rate. We shall assume that the velocity at which a snowplough is able to clear the road is inversely proportional to the depth of the accumulated snow. A snowplough leaves Kinlochewe at 09.00. It clears 16 miles of the road by 10.00 and another 8 miles by 11.00.

1. At what time did it start snowing?

2. A second snowplough starts at 10.00 and follows in the tracks of the first one. When does the second snowplough catch up with the first one?

3. As they meet, the driver of the first snowplough calls it a day, takes out cross–country skis and heads off into the surroundings. The second snowplough continues without a break to Applecross. When does the snowplough arrive there?

4. Discuss the major flaws of this model. How could you modify the velocity versus depth of snow model — by roughly remaining within the confines of inverse proportionality — such that for a clear road the velocity of the snowplough is v_0?

Scenario 2

Try to construct a more realistic model for the velocity of snowploughs. Assume that a snowplough travels at a constant velocity of 20 miles per hour when the road is clear of snow. Assume that as the snowplough encounters snow, the rate of change of its velocity with respect to the depth of snow is proportional to the depth of snow itself. Further, if the depth of snow reaches 100 cm the snowplough gets stuck. Using that model, find an expression for the velocity of the snowplough when the rate of snowfall is an arbitrary function of time.

1. Assume that the road from Kinlochewe to Applecross is uniformly covered by snow 50 cm deep. At 10.00 snow starts to fall at a constant rate of 10 cm/hour. A snowplough sets off from Kinlochewe immediately as the snow begins to fall. Will it reach Applecross or get stuck on the way? Give the time of arrival at Applecross or time/position of the snowplough as it gets stuck in the snow.

2. Sketch the graph of velocity of the snowplough versus depth of snow and compare it with the corresponding graphs from the original model in the first scenario and the modification as discussed in item 4.

Hint: in the modelling of this problem neglect acceleration and deceleration effects when the snowploughs start or stop.

A.4 FOXES AND HARES

Project summary:[1] This project deals with the solution of systems of differential equations of first order and an application to population dynamics.

Introduction

Rather than find a single function, say $y(x)$, from a single differential equation, systems of differential equations comprise a number of differential equations which must be solved simultaneously to obtain the solution for a number of functions, say $y_n(x), n = 1, 2, 3, ..., N$. The standard form of a linear system of first order differential equations with constant coefficients for N unknown functions $y_1(x), y_2(x), ..., y_N(x)$, is

$$
\begin{aligned}
y_1'(x) &= a_{11}y_1(x) + a_{12}y_2(x) + \ldots + a_{1N}y_N(x) + b_1(x) \\
y_2'(x) &= a_{21}y_1(x) + a_{22}y_2(x) + \ldots + a_{2N}y_N(x) + b_2(x) \\
&\vdots \qquad\qquad\qquad\qquad \vdots \\
y_n'(x) &= a_{n1}y_1(x) + a_{n2}y_2(x) + \ldots + a_{nN}y_N(x) + b_n(x) \\
&\vdots \qquad\qquad\qquad\qquad \vdots \\
y_N'(x) &= a_{N1}y_1(x) + a_{N2}y_2(x) + \ldots + a_{NN}y_N(x) + b_N(x)
\end{aligned}
\tag{A.13}
$$

or, in more compact form

$$
y_k'(x) = \sum_{j=1}^{N} a_{kj}\, y_j(x) + b_k(x), \quad (k = 1, 2, 3, .., N)
\tag{A.14}
$$

where a_{kj} are constants and $b_k(x)$ are given functions. To select a particular solution from the general solution these N differential equations must be supplemented by N linearly independent relations (the initial conditions).

Part 1: The case $N = 2$

Consider the case of $N = 2$, i.e., we have two equations for two unknown functions.

1. By differentiation and elimination show that the system of two equations is equivalent to one differential equation of second order for one of the two unknown functions. Provide the solution to that equation

---

[1]The project is a combination of material covered in Chapter 8 and Tutorial example T 8.2. Consequently the project — in its aim to introduce the concept of systems of differential equations of first order — is only pedagogically useful if it is tackled *before* the referenced book material is covered.

(schematically) and show how the second unknown function can be obtained once that solution is obtained. Can you see how, alternatively, any second order differential equation with constant coefficients can be written as a system of two first order differential equations?

2. Assume that $b_1(x) = b_{10}$ and $b_2(x) = b_{20}$ both reduce to constants. Find the **state of equilibrium** which is defined as

$$y'_k(x) = 0, \qquad k = 1, 2. \tag{A.15}$$

Find a substitution (based on the result for the equilibrium position) which transforms the inhomogeneous system into a homogeneous one. Obtain the particular solution (of the inhomogeneous system) which fulfils

$$y_k(0) = y_{k0}, \qquad k = 1, 2 \tag{A.16}$$

for two constants y_{10}, y_{20}.

3. Remain with the case $b_1(x) = b_{10}$ and $b_2(x) = b_{20}$. Try to solve the differential equations without the detour to second order equations. This consists of two steps because the general solution (as with individual differential equations) is a sum of the *complementary function* (i.e., a solution of the homogeneous equations) plus a *particular integral* (i.e., any solution of the inhomogeneous equations). To determine the complementary function assume that

$$y_1(x) = A_1 e^{\lambda x}, \qquad y_2(x) = A_2 e^{\lambda x} \tag{A.17}$$

where A_1, A_2 and λ are constants to be determined. The particular solution can be found by *guessing*. In this procedure do you see any parallels to the solution procedure for linear algebraic equations?

Part 2: Foxes and hares

Consider a somewhat primitive predator–prey model for a population of foxes and hares. Assume that the fox population $F(t)$ increases at a rate which is proportional to the size of the hare population $H(t)$ and is reduced by a mortality rate of a_F. Similarly, the hare population $H(t)$ declines at a rate proportional to the size of the fox population $F(t)$ but is sustained by a large birth rate a_H $(> a_F)$.

Develop the model differential equations which describe how the fox and hare populations change in time. Assuming that a_H and a_F are given, determine the two remaining proportionality constants from the observation that for some period now the fox and hare populations have been static at F_0 and H_0 respectively.

A sudden outbreak of *leveret myxomatosis* rapidly reduces the hare population to a fraction k of its previous level thus creating an ecological imbalance. Construct differential equations for $F(t)$ and $H(t)$ and hence deduce a second order differential equation for $F(t)$. Use this model to predict the future development of the fox and hare populations — with particular emphasis on the long–term behaviour.

Appraise critically these conclusions and speculate how the model might be modified to make it more realistic.

A.5 BEYOND THE LOGISTIC EQUATION

Project summary: This project examines mathematical models for population growth or decline.

Introduction

Many applications of mathematics, ranging from medicine to ecology to global economics, require the formulation of a model which predicts the future growth or decline of a species or a commodity. One of the most important models is provided by the **logistic equation** which assumes that the growth rate (of a species, say) depends on the population. This leads to the simple differential equation

$$\frac{dN}{dt} = r \left(1 - \frac{N}{K} \right) N \qquad (A.18)$$

where $N(t)$ is the population at time t while r and K are constants.[2]

Mathematical properties of the logistic equation

In the first part of this project the aim is to select reading material which will cover some of the simpler mathematical concepts associated with the logistic equation: *equilibrium points*, *critical points* and *equilibrium solutions*; *saturation levels* and the concept of *stability of solutions*; growth with and without *thresholds*. It shall be left to the judgement of the individual or the group tackling the project to decide upon a suitable cross–section of material for presentation.[3]

---

[2]The logistic equation, in slightly different notation, appeared in (2.9) of Section 2.2.

[3]A good starting point for a literature search is the book by J. D. Murray, *Mathematical Biology* (Springer Verlag, Berlin, 2nd edition, 1993). The reader should be aware that the topic of non–linear differential equations leads quickly into the complex realm of *chaos*.

Harvesting and the Schaefer equation

Based on the mathematical ideas developed in the previous subsection, a more sophisticated model of population dynamics will be investigated. In any economic scenario a renewable resource — such as a population of fish, for example — will be harvested. To model the harvesting process, the logistic equation has to be modified by introducing an additional term which models the depletion of the resource through external influences. In one such scenario the logistic equation is replaced by

$$\frac{dN}{dt} = r\left(1 - \frac{N}{K}\right)N - EN, \qquad (A.19)$$

a differential equation commonly known as the **Schaefer equation**. The new term $Y = EN$ is called the *yield*; it corresponds to the rate at which the resource is harvested (E is a positive constant with unit of time^{-1}).

1. Find the two equilibrium points in the case $E < r$.

2. Show that one of the equilibrium points is stable and the other one unstable.

3. Find the sustainable yield Y as a function of the effort E and draw the graph of this function (which is known as the yield–effort curve).

4. Find the maximum sustainable yield $Y_{\max}$, by determining the E which maximizes Y.

A.6 BOUNCING ABOUT

Project summary: In this project equations of motion of masses attached to springs are studied.

Introduction

Consider the following mechanical system: a mass m_1 is attached to a fixed point via a spring with spring constant k_1. A second mass m_2 is connected to the first mass by a spring of spring constant k_2 and the springs themselves are assumed to be massless. Define $u_1(t)$ and $u_2(t)$ as the displacement of the two masses from the position where the springs have their natural, unstretched length (we will assume here that all motions are restricted to one spatial dimension, the vertical one). Each mass is subject to the force of gravity and a restoring force due to the springs. The restoring force of a spring is proportional to its displacement from its natural length (the proportionality constant being the spring constant) and directed in the opposite direction.

Model equations and tasks

Under the assumptions as outlined above show that the equations of motion for the two masses are given by

$$m_1 \ddot{u}_1 = -(k_1 + k_2)u_1 + k_2 u_2 + m_1 g$$
$$m_2 \ddot{u}_2 = -k_2 u_2 + k_2 u_1 + m_2 g \tag{A.20}$$

where $g = 9.81$ m/s^2 is the gravitational acceleration. Address the following tasks.

1. Find the equilibrium positions $u_{1\text{eq}}$ and $u_{2\text{eq}}$.

2. Define new dependent functions

$$v_1(t) = u_1(t) - u_{1\text{eq}}, \qquad v_2(t) = u_2(t) - u_{2\text{eq}}$$

and rewrite the above two differential equations in terms of v_1 and v_2.

3. Set $m_1 = m_2 = 1$, $k_1 = 3$ and $k_2 = 2$. Initially, the two springs are at rest but displaced from their equilibrium positions. Solve the initial value problem for $v_1(t)$ and $v_2(t)$:

$$v_1(0) = v_{10}, \quad \dot{v}_1(0) = 0, \quad v_2(0) = v_{20}, \quad \dot{v}_2(0) = 0. \tag{A.21}$$

4. Discuss the solution. Notice that it contains two distinct modes of vibration. For which combination of the constants v_{10}, v_{20} does the solution represent a vibration whereby the masses move

   (a) in phase, i.e. both moving up and down together?

   (b) out of phase, i.e. one moving down while the other is moving up (and vice versa)?

   What are the respective frequencies of these vibrations?

5. Try to generalize this simple mechanical problem of 2 masses on 2 springs to one of N masses $m_n, n = 1, 2, ..., N$ on N springs with spring constants $k_n, n = 1, 2, ..., N$.

   (a) Formulate the system of model differential equations.

   (b) Draw up a scheme of how you would go about solving them.

6. Discuss some general limitations of the model.

Appendix B

Extended Tutorial Solutions

TUTORIAL EXAMPLES 1

T 1.1   This is a *linear* differential equation. Its integrating factor is

$$\mu(x) = e^{\int (-1)dx} = e^{-x}.$$

Therefore,

$$e^{-x}y(x) = \int e^{-x}\sin 2x\, dx.$$

The integral can be evaluated using integration by parts twice. One obtains

$$e^{-x}y(x) = e^{-x}\left(-\frac{1}{5}\sin 2x - \frac{2}{5}\cos 2x\right) + C$$

and the general solution is

$$y = -\frac{1}{5}\sin 2x - \frac{2}{5}\cos 2x + Ce^{x}.$$

T 1.2   The differential equation can be rewritten in the form

$$y' + \frac{1}{x}y = -\frac{1}{x^2}.$$

It has an integrating factor

$$\mu(x) = e^{\int (1/x)dx} = e^{\ln x} = x$$

and therefore

$$xy(x) = \int x\left(-\frac{1}{x^2}\right)dx = -\ln x + C.$$

The general solution is then

$$y(x) = \frac{C - \ln x}{x}.$$

T 1.3 This differential equation is *separable*. Thus

$$\frac{dy}{1 - y^2} = -\tan x \, dx$$

$$\longrightarrow \quad \frac{1}{2}\left(\frac{dy}{1+y} + \frac{dy}{1-y}\right) = -\tan x \, dx$$

$$\longrightarrow \quad \frac{1}{2}\Big[\ln|1+y| - \ln|1-y|\Big] = \ln|\cos x| + C$$

$$\longrightarrow \quad \left|\frac{1+y}{1-y}\right|^{1/2} = D|\cos x| \qquad \longrightarrow \qquad \frac{1+y}{1-y} = E\cos^2 x\,.$$

The general solution is therefore

$$y(x) = \frac{\cos^2 x - A}{\cos^2 x + A}\,.$$

T 1.4 This first order differential equation is *homogeneous*. Using the substitution $y(x) = xu(x)$, $y'(x) = xu'(x) + u(x)$, one gets

$$(1 - 6u^2)(xu' + u) = 4u(1 + 3u^2) \qquad \longrightarrow \qquad (1 - 6u^2)xu' = 3u(1 + 6u^2)\,.$$

This equation is now *separable*. Therefore,

$$\frac{(1 - 6u^2)}{(1 + 6u^2)u} \, du = \frac{3}{x} \, dx \quad \longrightarrow \quad \left(\frac{1}{u} - \frac{12u}{1 + 6u^2}\right) du = \frac{3}{x} \, dx$$

$$\longrightarrow \quad \ln u - \ln(1 + 6u^2) = 3\ln x + C \quad \longrightarrow \quad \frac{u}{1 + 6u^2} = Dx^3\,.$$

Returning to the original dependent variable y, the general solution is therefore

$$\frac{y/x}{1 + 6y^2/x^2} = Dx^3 \quad \longrightarrow \quad \frac{y}{x^2 + 6y^2} = Dx^2\,.$$

T 1.5 This differential equation is of the form $P + Qy' = 0$ and because $\partial P/\partial y = 4x$, $\partial Q/\partial x = 4x$, it is *exact*. Thus

$$u(x,y) = \int P(x,y) \, dx = \int (3x^2 + 4xy) \, dx = x^3 + 2x^2y + \phi(y)$$

This leads, finally, to the general solution

$$y(x) = x + \frac{a}{2x + D(x^2 + a)^{1/2}}\,.$$

T 1.9   We divide the equation by $x^2 + 1$ and get

$$y' - \frac{x}{x^2 + 1}\,y = x\,.$$

This linear equation has an integrating factor

$$\mu(x) = e^{\int[-x/(x^2+1)]\,dx} = e^{-(1/2)\ln(x^2+1)} = (x^2 + 1)^{-1/2}\,.$$

Therefore

$$(x^2 + 1)^{-1/2}\,y(x) = \int^x (s^2 + 1)^{-1/2}\,s\,ds = (x^2 + 1)^{1/2} + C$$

with the general solution in final form as

$$y(x) = C(x^2 + 1)^{1/2} + x^2 + 1\,.$$

T 1.10   An integrating factor is

$$\mu(x) = e^{\int 2x\,dx} = e^{x^2}\,.$$

Then,

$$e^{x^2} y(x) = \int e^{x^2} e^{-x^2}\,dx = \int dx = x + C$$

and thus

$$y(x) = (x + C)\,e^{-x^2}\,.$$

Substitution of the initial condition gives $C = 1$ and so the particular solution is

$$y(x) = (x + 1)\,e^{-x^2}\,.$$

T 1.11   An integrating factor of this linear equation is

$$\mu(x) = e^{\int(2/x)\,dx} = e^{2\ln x} = x^2\,.$$

Hence

$$x^2 y(x) = \int x^2\,\frac{\cos x}{x^2}\,dx = \sin x + C\,.$$

Consequently, the general solution is

$$y(x) = \frac{C + \sin x}{x^2}.$$

The initial condition requires $C = 0$, so that one gets as particular solution

$$y(x) = \frac{\sin x}{x^2}.$$

T 1.12 The differential equation is of *Bernoulli* type. Using the substitution

$$z(x) = y^{-4}(x) \qquad \longrightarrow \qquad z' = -4y^{-5}y',$$

one gets after algebraic rearrangement

$$z' + 2xz = -4x$$

with an integrating factor

$$\mu(x) = e^{\int 2x\, dx} = e^{x^2}.$$

Thus

$$e^{x^2} z(x) = \int e^{x^2}(-4x)\, dx = -2\, e^{x^2} + C.$$

Hence

$$z(x) = -2 + Ce^{-x^2} \qquad \longrightarrow \qquad y(x) = \frac{1}{(-2 + Ce^{-x^2})^{1/4}}.$$

Substituting the initial condition gives

$$a = \frac{1}{(-2 + C)^{1/4}} \qquad \longrightarrow \qquad C = 2 + a^{-4}$$

and the particular solution

$$y(x) = \frac{1}{\left[-2 + (2 + a^{-4})\, e^{-x^2}\right]^{1/4}}.$$

T 1.13 This differential equation is of *homogeneous* type. Substitution of $y(x) = xu(x)$, $y' = xu' + u$ gives

$$-x(1 + au^3)u' = au^4 \qquad \longrightarrow \qquad \left(\frac{1}{u^4} + \frac{a}{u}\right) du = -\frac{a}{x}\, dx$$

$$\longrightarrow \qquad -\frac{1}{3au^3} + \ln u = -\ln x + C \qquad \longrightarrow \qquad ue^{-(1/3au^3)} = \frac{C}{x}.$$

Thus

$$y = Ce^{(x^3/3ay^3)}$$

is the general solution. The initial condition gives $C = 1$ leading to the particular solution

$$y = e^{(x^3/3ay^3)}.$$

T 1.14 An integrating factor is

$$\mu(x) = e^{\int (-2)dx} = e^{-2x}.$$

Thus

$$e^{-2x}y(x) = \int^x e^{-2s}s^2 e^{2s}\, ds = \int^x s^2\, ds = \frac{x^3}{3} + C.$$

The general solution is

$$y(x) = \left(\frac{x^3}{3} + C\right)e^{2x}.$$

Substitution of the initial condition yields $C = 0$ and the particular solution

$$y(x) = \frac{x^3 e^{2x}}{3}.$$

T 1.15 After rewriting the differential equation in the form

$$2x - y + (2y - x)y' = 0,$$

a straightforward check shows that this equation is exact. Consequently,

$$u(x) = \int (2x - y)\, dx = x^2 - xy + \phi(y)$$

and

$$\frac{\partial u}{\partial y} = -x + \phi'(y) = 2y - x.$$

From the last equation, we see that $\phi(y) = y^2$ which means that the general solution is $x^2 - xy + y^2 = C$. The initial condition determines $C = 7$ so that the particular solution is

$$x^2 - xy + y^2 - 7 = 0.$$

In explicit form, the particular solution is

$$y(x) = \frac{1}{2}\left(x + \sqrt{28 - 3x^2}\right).$$

Note that of the two possibilities, the plus sign was chosen for the square root to satisfy the initial condition. A real solution is only possible if $28 - 3x^2 \geq 0$ so that, with the initial value prescribed at $x = 1$, the range of validity of the solution is

$$1 \leq x \leq \sqrt{\frac{28}{3}}\,.$$

TUTORIAL EXAMPLES 2

T 2.1   Let $M(t)$ be the mass of ^{14}C at time t, then radioactive decay is modelled by the differential equation

$$\frac{dM(t)}{dt} = -\alpha M(t)$$

where α is a proportionality constant. The solution of this differential equation fulfilling the initial condition $M(0) = M_0$ is

$$M(t) = M_0 e^{-\alpha t}\,.$$

It then follows from the definition of the half–life, τ, that $M(\tau) = M_0/2$ and therefore

$$\frac{M_0}{2} = M_0 e^{-\alpha \tau} \quad \longrightarrow \quad \tau = \frac{1}{\alpha}\ln 2 \quad \longrightarrow \quad \alpha = \frac{1}{\tau}\ln 2$$

and thus

$$M(t) = M_0 e^{-(t/\tau)\ln 2} = M_0\, 2^{-t/\tau}\,.$$

Using the provided information, $\tau = 5,580$ years and $M(t^*) = 0.3M_0$, t^* satisfies

$$0.3M_0 = M_0\left(\frac{1}{2}\right)^{t^*/5580}$$

and has value

$$t^* = 5580\,\frac{\ln 0.3}{\ln 0.5} \approx 9692\ \text{years}\,.$$

T 2.2   The environment temperature is given by $S(t) = S_0 + b\sin \omega t$ where S_0, b and ω are constants. The formula for the object's temperature yields (here we have $t_0 = 0$)

$$\begin{aligned} T(t) &= ke^{-kt}\int e^{kt} S(t)\,dt \\ &= S_0 + \frac{bk}{k^2 + \omega^2}(k\sin \omega t - \omega \cos \omega t) + Ce^{-kt}\,. \end{aligned}$$

The initial condition $T(0) = T_0$ gives $C = T_0 + S_0 + bk\omega/(k^2 + \omega^2)$ and so the particular solution is

$$T(t) = S_0 + \left(T_0 - S_0 + \frac{bk\omega}{k^2 + \omega^2}\right)e^{-kt} + \frac{bk}{k^2 + \omega^2}(k\sin\omega t - \omega\cos\omega t).$$

This formula may be interpreted as follows: the first two terms on the right–hand side provide exponential cooling to the temperature S_0 whereas the third term is the temperature variation imposed by the forcing term of the environmental temperature.

Special cases. If $\omega/k << 1$, then we have

$$T = S_0 + \left(T_0 - S_0 + \frac{b(\omega/k)}{1 + (\omega/k)^2}\right)e^{-kt} + \frac{b}{1 + (\omega/k)^2}\left(\sin\omega t - \frac{\omega}{k}\cos\omega t\right)$$

and thus

$$T(t) \approx S_0 + (T_0 - S_0)e^{-kt} + b\sin\omega t.$$

In this limit the temperature cools exponentially to the temperature S_0, and there is an *instantaneous* response to the trigonometric variation as given by the periodic component of the solution.

Analogously, when $k/\omega << 1$ we obtain

$$T(t) \approx S_0 + (T_0 - S_0)e^{-kt} - \frac{k}{\omega}\cos\omega t,$$

the last term on the right–hand side showing a delayed response of the system proportional to k/ω.

Large time. In the limit $t \to \infty$, we can see that as $e^{-kt} \to 0$ (for $k > 0$),

$$\lim_{t\to\infty}\left[T - \frac{bk}{k^2 + \omega^2}(k\sin\omega t - \omega\cos\omega t)\right] = S_0.$$

A brief word about limitations of such simple models. In reality, it takes a finite time for the effects of a source such as $S(t)$ to be felt at any given point. The cooling law, for example, makes no concession to spatial variation of temperature. That is clearly unphysical. A better model would require one to assume that the temperature is not just a function of time as in $T(t)$ but of time and space as in $T(x, y, z, t)$. This leads to the heat conduction equation in which spatial variations of temperature are often modelled by $\partial^2 T/\partial x^2 + \partial^2 T/\partial y^2 + \partial^2 T/\partial z^2$.

T 2.3   The solution of the logistic equation is

$$P(t) = \frac{MP_0}{P_0 + (M - P_0)e^{-rMt}}.$$

We have $M = 2550$ (population in thousands) and the census data

$$
\begin{array}{lll}
1800 & t_0 = 0 & P_0 = 739 \\
1850 & t_1 = 50 & P_1 = 1376 \\
2000 & t_2 = 200 & P_2 = \ ?
\end{array}
$$

One needs to find r from t_1 and P_1 (t_0, P_0 are already incorporated into the formula for $P(t)$ by the setting of $P = P_0$ at $t = 0$). Rearrangement of the expression for $P(t)$ leads to

$$
r = \frac{1}{Mt_1} \ln \frac{P_1(M - P_0)}{P_0(M - P_1)} = 8.27529 \times 10^{-6} \, .
$$

Therefore

$$
P(200) = \frac{2550 \times 739}{739 + 1811 \, e^{-4.2204}} = 2461.375 \, .
$$

The male population of Scotland is therefore estimated as 2,461,000 in the year 2000.

T 2.4 Let $\phi = \ln W$ then $\dot{\phi} = a - b\phi$ with solution

$$
\phi = \phi_0 e^{-bt} + \frac{a}{b}\left(1 - e^{-bt}\right), \qquad \phi_0 = \ln W_0 \, .
$$

The given information yields the two relations

$$
\ln W_k - \frac{a}{b} = \left(\phi_0 - \frac{a}{b}\right) e^{-bt_k} , \qquad k = 1, 2 \, .
$$

Therefore

$$
\left(\ln W_1 - \frac{a}{b}\right)^{t_2} = \left(\phi_0 - \frac{a}{b}\right)^{t_2} e^{-bt_1 t_2} ,
$$

$$
\left(\ln W_2 - \frac{a}{b}\right)^{t_1} = \left(\phi_0 - \frac{a}{b}\right)^{t_1} e^{-bt_1 t_2} ,
$$

which leads to

$$
\left(\ln W_1 - \frac{a}{b}\right)^{t_2} = \left(\phi_0 - \frac{a}{b}\right)^{t_2 - t_1} \left(\ln W_2 - \frac{a}{b}\right)^{t_1} .
$$

This is generally a transcendental equation for a/b. However, if $t_2 = 2t_1$ then

$$
\left(\ln W_1 - \frac{a}{b}\right)^2 = \left(\phi_0 - \frac{a}{b}\right)\left(\ln W_2 - \frac{a}{b}\right)
$$

with solution

$$
\frac{a}{b} = \frac{(\ln W_1)^2 - (\ln W_0)(\ln W_2)}{\ln[W_1^2/(W_0 W_2)]} .
$$

We see that $v(t) \to 0$ as $t \to \infty$.

T 2.6 We have the differential equation

$$m \frac{dv}{dt} = -\frac{k}{v^2}$$

and thus

$$\int v^2 \, dv = \int -\frac{k}{m} \, dt \qquad \longrightarrow \qquad \frac{v^3}{3} = -\frac{k}{m} t + C \, .$$

The initial condition is $v(0) = v_0$, so that $C = v_0^3/3$. The particular solution is therefore

$$v^3 = v_0^3 - \frac{3kt}{m} \qquad \longrightarrow \qquad v(t) = \left(v_0^3 - \frac{3kt}{m} \right)^{1/3} \, .$$

The time t^* satisfies $v(t^*) = 0$ and so $t^* = mv_0^3/3k$. The total distance travelled is therefore

$$
\begin{aligned}
d &= \int_0^{t^*} v(t) \, dt = \int_0^{t^*} \left(v_0^3 - \frac{3kt}{m} \right)^{1/3} dt \\
&= \left[\left(v_0^3 - \frac{3kt}{m} \right)^{4/3} \frac{3}{4} \left(-\frac{m}{3k} \right) \right]_0^{t^*} \\
&= -\frac{m}{4k} \left[\left(v_0^3 - \frac{3kt^*}{m} \right)^{4/3} - v_0^4 \right] = \frac{mv_0^4}{4k} \, .
\end{aligned}
$$

To determine the initial velocity of the puck from the given experimental values, we note that neither k nor m are known (but they only occur as ratio k/m in all formulas). However, from the expression for t^*, we can express $m/k = 3t^*/v_0^3$. Substituting this result into the formula for d gives

$$d = \frac{3v_0 t^*}{4} \qquad \longrightarrow \qquad v_0 = \frac{4d}{3t^*} \, .$$

Finally, we can now set $d = 24$ m and $t^* = 8$ s and get $v_0 = 4$ m/s as the initial velocity of the ice–hockey puck.

T 2.7 Let V be the constant volume of the reservoir, I the constant inflow (which equals the outflow) and $P(t)$ the volume of pollutant in the reservoir. Let $y(t) = P(t)/V$ be the volume concentration of pollutant in the reservoir and $q(t)$ be the volume concentration of pollutant in the inflow. The differential equation modelling the volume of pollutant in the reservoir is

$$\frac{dP}{dt} = q(t)I - y(t)I$$

(in words: the rate of change of pollutant equals inflow of pollutant minus outflow). Therefore

$$V \frac{dy(t)}{dt} = q(t)I - y(t)I \qquad \longrightarrow \qquad \dot{y} + \frac{I}{V} y = \frac{I}{V} q .$$

This is a *linear* differential equation with the general solution

$$y(t) = e^{-(I/V)t} \frac{I}{V} \int^t e^{(I/V)s} q(s) \, ds .$$

The initial condition is $y(0) = y_0$.
$q(t) \equiv 0$. The integral becomes trivial:

$$y(t) = y_0 \, e^{-(I/V)t} ,$$

so the pollutant concentration in the reservoir decays exponentially to zero.
$q(t) = q_0$. The particular solution is

$$y(t) = q_0 + (y_0 - q_0) \, e^{-(I/V)t} .$$

The pollutant concentration in the reservoir approaches the pollutant concentration in the inflow exponentially (increasing or decreasing, depending on $y_0 < q_0$ or $y_0 > q_0$).
    Next, we set $V = 2 \, 10^6$ m^3 and $I = 2$ m^3/s and let

$$q(t) = \begin{cases} 0 & t \leq 0 \\ 10^{-4} & 0 < t \leq 10^6 \\ 0 & t > 10^6 . \end{cases}$$

All the ingredients for the solution are already available.

- $t \leq 0$. For $q(t) \equiv 0$ one gets $y = y_0 \, e^{-10^{-6}t}$ but because $y \equiv 0$ for $t \leq 0$ one must have $y(0) = 0$ and therefore

$$y(t) \equiv 0 , \qquad t \leq 0 .$$

- $0 < t \leq 10^6$. With $q_0 = 10^{-4}$ and $y(0) = 0$ one gets

$$y(t) = 10^{-4} \left(1 - e^{-10^{-6}t} \right) , \qquad 0 < t \leq 10^6 ,$$

  in particular, one finds the value $y(10^6) = 10^{-4}(1 - e^{-1})$.

- $t > 10^6$. Again $q(t) \equiv 0$ but the 'initial' value is $y(10^6) = 10^{-4}(1-e^{-1})$. Adjustment of the integration constant leads to the particular solution

$$y(t) = 10^{-4}(1 - e^{-1})e^{1-10^{-6}t} , \qquad t > 10^6 .$$

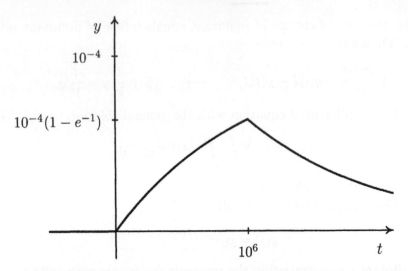

Figure B.1: Volume concentration y of pollutant in the reservoir as a function of time t as in Tutorial example T 2.7.

A graph of the solution is shown in Figure B.1.

T 2.8   The solution proceeds as in Section 2.4 but now for the case $w \leq v$. Treating first $w < v$, the solution is again

$$y = \frac{a}{2}\left[\frac{(x/a)^{v/w+1}}{1+v/w} - \frac{(x/a)^{v/w-1}}{1-v/w}\right] + \frac{avw}{w^2 - v^2}.$$

But because $1 - v/w < 0$, the line $x = 0$ is an asymptote which geometrically reflects the fact that the fox can never reach the rabbit.

For $w = v$ one needs to start from

$$y' = \frac{1}{2}\left(\frac{x}{a} - \frac{a}{x}\right).$$

Integration yields

$$y = \frac{1}{2}\left(\frac{x^2}{2a} - a\ln x\right) + C.$$

The initial condition $y(a) = 0$ gives $C = (a/4)(2\ln a - 1)$ and thus

$$y = \frac{1}{2}\left(\frac{x^2 - a^2}{2a} - a\ln\frac{x}{a}\right).$$

Once again, $x = 0$ is an asymptote. The distance between the fox and the rabbit is

$$d^2 \;=\; x^2 + (y - vt)^2 = x^2\left[1 + \left(\frac{y - vt}{x}\right)^2\right] = x^2\left(1 + y'^2\right)$$

$$= x^2 \left[1 + \frac{1}{4} \left(\frac{x}{a} - \frac{a}{x} \right)^2 \right] = \frac{x^2}{4} \left(\frac{x}{a} + \frac{a}{x} \right)^2 .$$

Therefore,

$$d = \frac{1}{2a} \left(x^2 + a^2 \right) \qquad \longrightarrow \qquad \lim_{x \to 0} d = \frac{a}{2} .$$

T 2.9   Let the research vessel R travel for 3 km towards the place where the whale W was spotted, then the whale's position is located somewhere on a circle of radius 1 km centred at the point where the whale was initially seen. Take this moment as the initial time. Choose circular polar coordinates (r, θ) with the origin at the initial position of the whale and let therefore $r = r(\theta)$ describe the path of R. Then, the distances travelled after a certain time t are

$$r - 1 \qquad\qquad\qquad\qquad \text{for W} ,$$

$$\int_0^{s(\theta)} ds = \int_0^\theta \sqrt{ \left(\frac{dr}{d\theta} \right)^2 + r^2 } \, d\theta \qquad \text{for R} .$$

Let v_R and v_W denote the velocities of R and W respectively. It is known that $v_R = 3v_W$. For their paths to cross, R and W must be at the same place at the same time, say t^*. Therefore

$$t^* = \frac{r-1}{v_W} = \frac{1}{v_R} \int_0^\theta \left(r^2 + r'^2 \right)^{1/2} d\theta$$

or

$$3(r-1) = \int_0^\theta \left(r^2 + r'^2 \right)^{1/2} d\theta$$

where $r' = dr/d\theta$. Differentiation with respect to θ gives

$$3r' = \left(r^2 + r'^2 \right)^{1/2} \qquad \longrightarrow \qquad r' = \frac{\pm r}{\sqrt{8}} \qquad \longrightarrow \qquad r = Ce^{\pm\theta/\sqrt{8}} .$$

With the initial condition $r = 1$ for $\theta = 0$, particular solutions emerge as the spirals $r = e^{\pm\theta/\sqrt{8}}$. It is easy to convince oneself that of the two choices only the one with positive exponent is relevant. (An alternative way to set up the solution is to let R travel for 6 km in the direction of W. Then again a spiral arises as the solution: $r = 2e^{(\theta-\pi)/\sqrt{8}}$.)

T 2.10   The voltage balance in the described circuit is $U(t) = U_C + U_R$ where U_C and U_R are the voltage drops on the capacitor and resistor respectively. Using[2] $U_R = RI$, $U_C = Q/C$ and $I = dQ/dt$ one obtains the

---

[2] We remind that reader that in examples dealing with electrical circuits, C denotes the capacitance and not an integration constant.

differential equation

$$R\dot{Q} + \frac{Q}{C} = U\,.$$

The general solution is

$$Q(t) = \frac{1}{R}\,e^{-t/RC} \int^{t} e^{s/RC}\,U(s)\,ds\,.$$

1. $U(t) \equiv 0, Q(t_0) = Q_0$.

   In this case $Q(t) = Ae^{-t/RC}$ for some integration constant A. Substitution of the initial condition leads to the particular solution

   $$Q(t) = Q_0\,e^{-(t-t_0)/RC}\,.$$

   Interpretation: The capacitor discharges through the resistor and its charge decreases exponentially. The current flowing through the circuit is

   $$I(t) = \frac{dQ(t)}{dt} = -\frac{Q_0}{RC}\,e^{-(t-t_0)/RC}\,.$$

2. $U(t) = U_0, Q(t_0) = 0$.

   In this case, integration produces

   $$Q(t) = \frac{U_0}{R}\,e^{-t/RC}\left(RCe^{t/RC} + A\right) = U_0C + B\,e^{-t/RC}\,.$$

   The initial condition yields $B = -U_0Ce^{t_0/RC}$ for the integration constant, leading to the particular solution

   $$Q(t) = U_0C\left[1 - e^{-(t-t_0)/RC}\right]\,.$$

   Interpretation: From being initially uncharged, the capacitor in the circuit is charged through the resistor. It attains the maximum charge $Q_{max} = U_0C$ (in the limit $t \to \infty$) and therefore

   $$\frac{Q_{max}}{2} = Q_{max}\left[1 - e^{-(t^*-t_0)/RC}\right]$$

   determines the time t^* when the capacitor is charged to half its maximum value. It follows that

   $$e^{-(t^*-t_0)/RC} = \frac{1}{2} \qquad \longrightarrow \qquad t^* = t_0 + RC\ln 2\,.$$

   (Compare this conceptually with the half–life in radioactivity.)

3. $U(t) = U_0 \sin \omega t$.

$$U_C(t) = \frac{Q(t)}{C} = \frac{U_0}{RC} e^{-t/RC} \int e^{t/RC} \sin \omega t \, dt .$$

Integration by parts now gives the solution

$$U_C(t) = \frac{U_0}{1 + (RC\omega)^2} (\sin \omega t - RC\omega \cos \omega t) + A e^{-t/RC} .$$

As $t \to \infty$, we obtain the limit

$$\lim_{t \to \infty} \left[U_C(t) - \frac{U_0}{1 + (RC\omega)^2} (\sin \omega t - RC\omega \cos \omega t) \right] = 0 .$$

TUTORIAL EXAMPLES 3

T 3.1   The key step in the calculation is the evaluation of

$$\Phi(\lambda, t) = \int_0^t e^{\lambda s} f(t - s) \, ds .$$

The auxiliary equation is $\lambda^2 + 3\lambda + 2 = 0$ with roots $\lambda_1 = -1$ and $\lambda_2 = -2$.
We need to calculate

$$
\begin{aligned}
\Phi(-1, t) &= \int_0^t e^{-s} (t - s) e^{-2(t-s)} \, ds = \int_0^t e^s (t - s) e^{-2t} \, ds \\
&= e^{-2t} \left[e^s (t - s) + e^s \right]_0^t = e^{-2t} \left(-t + e^t - 1 \right) , \\
\Phi(-2, t) &= \int_0^t e^{-2s} (t - s) e^{-2(t-s)} \, ds = \int_0^t (t - s) e^{-2t} \, ds \\
&= e^{-2t} \left[ts - \frac{s^2}{2} \right]_0^t = \frac{t^2}{2} e^{-2t} .
\end{aligned}
$$

Therefore the general solution is

$$
\begin{aligned}
y(t) &= A e^{-t} + B e^{-2t} + \frac{\Phi(-1, t) - \Phi(-2, t)}{-1 - (-2)} \\
&= C e^{-t} + D e^{-2t} - \left(t + \frac{t^2}{2} \right) e^{-2t} .
\end{aligned}
$$

T 3.2   The auxiliary equation is $\lambda^2 + 2\lambda + 5 = 0$ with roots $\lambda_{1,2} = -1 \pm 2i$. We compute

$$
\begin{aligned}
\Phi(-1+2i,t) &= \int_0^t e^{(-1+2i)s} \cos 2(t-s)\, ds \\
&= \frac{1}{2} \int_0^t e^{(-1+2i)s} \left[e^{2(t-s)i} + e^{-2(t-s)i} \right] ds \\
&= \frac{1}{2} \int_0^t \left(e^{2it} e^{-s} + e^{-2it} e^{4is-s} \right) ds \\
&= \frac{1}{2} \left(1 - e^{-t} \right) e^{2it} - \frac{4i+1}{34} \left(e^{2it-t} - e^{-2it} \right).
\end{aligned}
$$

Simple substitution of $i \to -i$ gives

$$
\Phi(-1-2i,t) = \frac{1}{2} \left(1 - e^{-t} \right) e^{-2it} - \frac{(-4i+1)}{34} \left(e^{-2it-t} - e^{2it} \right)
$$

and so

$$
\begin{aligned}
\frac{\Phi(-1+2i,t) - \Phi(-1-2i,t)}{-1+2i-(-1-2i)} &= \frac{1}{4} \left(1 - e^{-t} \right) \sin 2t \\
&\quad - \frac{1}{34} \left(-2\cos 2t + 2e^{-t} \cos 2t \right) - \frac{1}{68} \left(\sin 2t + e^{-t} \sin 2t \right) \\
&= \frac{4}{17} \sin 2t + \frac{1}{17} \cos 2t + \dots ,
\end{aligned}
$$

the dots indicating terms that are irrelevant as they are contained already in the complementary function. Therefore the general solution is

$$
y(t) = A e^{-t} \cos 2t + B e^{-t} \sin 2t + \frac{1}{17} \left(\cos 2t + 4 \sin 2t \right).
$$

T 3.3   As before, the auxiliary equation is $\lambda^2 + 2\lambda + 5 = 0$ with roots $\lambda_{1,2} = -1 \pm 2i$. Integration yields

$$
\begin{aligned}
\Phi(-1+2i,t) &= \int_0^t e^{(-1+2i)s} \frac{e^{-(t-s)}}{2} \left[e^{2i(t-s)} + e^{-2i(t-s)} \right] ds \\
&= \frac{e^{-t}}{2} \int_0^t \left(e^{2it} + e^{-2it} e^{4is} \right) ds \\
&= \frac{t}{2} e^{-t} e^{2it} + \frac{1}{4} e^{-t} \sin 2t .
\end{aligned}
$$

Analogously,

$$
\Phi(-1-2i,t) = \frac{t}{2} e^{-t} e^{-2it} + \frac{1}{4} e^{-t} \sin 2t .
$$

Finally,

$$\frac{\Phi(-1+2i,t)-\Phi(-1-2i,t)}{-1+2i-(-1-2i)}=\frac{t}{4}e^{-t}\sin 2t$$

so that the general solution is given by

$$y(t)=Ae^{-t}\cos 2t+Be^{-t}\sin 2t+\frac{t}{4}e^{-t}\sin 2t\,.$$

T 3.4   The auxiliary equation factorizes as $(\lambda+2)(\lambda+3)=0$ and therefore $y(t)=Ae^{-2t}+Be^{-3t}$. From the initial conditions

$$y(0)=1\ \longrightarrow\ A+B=1,\qquad \dot{y}(0)=a\ \longrightarrow\ -2A-3B=a\,.$$

Therefore $A=3+a$, $B=-a-2$ and

$$\begin{aligned}
y(t)\ &=\ (3+a)e^{-2t}-(a+2)e^{-3t}=e^{-3t}\Big[(a+3)e^{t}-(a+2)\Big]\\
&=\ e^{-3t}\Big[(a+3)(e^{t}-1)+1\Big]
\end{aligned}$$

is the particular solution. It follows that

$$a+3\geq 0\ \longrightarrow\ y(t)\geq 0,\qquad a+3<0\ \longrightarrow\ y(t)<0$$

(the latter relation applies for suitably large t). Therefore, $y(t)\geq 0$ *if and only if* $a\geq -3$.

T 3.5   The auxiliary equation is $\lambda^2+1=0$ with roots $\lambda_{1,2}=\pm i$. Thus

$$\begin{aligned}
\Phi(\pm i,t)\ &=\ \int_0^t e^{\pm is}\tan(t-s)\,ds=\int_0^t \tan\tau\, e^{\pm i(t-\tau)}\,d\tau\\
&=\ e^{\pm it}\int_0^t \frac{\sin\tau}{\cos\tau}\,(\cos\tau\mp i\sin\tau)\,d\tau\\
&=\ e^{\pm it}\int_0^t \sin\tau\mp\left(\frac{1-\cos^2\tau}{\cos\tau}\right)\,d\tau\\
&=\ e^{\pm it}\Big[-\cos\tau\mp i\ln|\sec\tau+\tan\tau|\pm i\sin\tau\Big]_0^t\\
&=\ e^{\pm it}\Big[-\cos t\mp i\ln|\sec t+\tan t|\pm i\sin t+1\Big]\\
&=\ e^{\pm it}\Big[1-e^{\mp it}\mp i\ln|\sec t+\tan t|\Big]
\end{aligned}$$

where $\sec t=1/\cos t$. We need to construct

$$\frac{\Phi(i,t)-\Phi(-i,t)}{i-(-i)}=\sin t-\cos t\,\ln|\sec t+\tan t|\,.$$

The general solution can then be written as

$$y(t) = A \sin t + B \cos t - \cos t \ln|\sec t + \tan t|.$$

The initial conditions give

$$y(0) = 1 \quad \longrightarrow \quad B = 1, \qquad \dot{y}(0) = 1 \quad \longrightarrow \quad A = 2$$

leading to the particular solution

$$y(t) = 2 \sin t + \cos t - \cos t \ln|\sec t + \tan t|.$$

TUTORIAL EXAMPLES 4

T 4.1   The complementary function is $y_c(x) = C_1 e^{-x} + C_2 e^{-2x}$. The choice $y_p(x) = (Ax + B)e^{-2x}$ will obviously not work because e^{-2x} is a solution of the homogeneous equation. Using Table 4.1 we have

$$y_p(x) = x(Ax + B)e^{-2x}$$

and therefore

$$y_p'(x) = \left[-2Ax^2 + (2A - 2B)x + B\right]e^{-2x},$$
$$y_p''(x) = \left[4Ax^2 + (4B - 8A)x + 2A - 4B\right]e^{-2x}.$$

Substitution into the differential equation and collection of terms leads to

$$[(-2A)x + 2A - B]e^{-2x} = xe^{-2x}.$$

It follows by comparison of coefficients that $A = -1/2$ and $B = 2A = -1$. Therefore the particular integral is

$$y_p(x) = \left(-\frac{x^2}{2} - x\right)e^{-2x}.$$

T 4.2   The complementary function is $y_c(x) = e^{-x}(A \cos 2x + B \sin 2x)$. The particular integral has form

$$y_p(x) = \alpha \cos 2x + \beta \sin 2x.$$

Substitution of y_p, y_p' and y_p'' into the differential equation and comparison of coefficients gives the linear system $\alpha + 4\beta = 1$ and $-4\alpha + \beta = 0$ with the solution $\alpha = 1/17$, $\beta = 4/17$. Therefore, the particular integral is

$$y_p(x) = \frac{1}{17}(\cos 2x + 4 \sin 2x).$$

T 4.3 The complementary function is the same as in the previous problem. For the particular integral we now make the choice

$$y_p(x) = x\,e^{-x}\,(\alpha \cos 2x + \beta \sin 2x) \ .$$

Again, after substitution of y_p, y_p' and y_p'' into the differential equation and comparison of coefficients we obtain $-4\alpha = 0$ and $4\beta = 1$. Therefore, the particular integral is

$$y_p(x) = \frac{1}{4}\,xe^{-x}\sin 2x \ .$$

T 4.4 The auxiliary equation is $\lambda^2 - 5\lambda + 6 = 0$ with roots $\lambda_1 = 2$, $\lambda_2 = 3$. Therefore the complementary function is $y_c(x) = C_1 e^{2x} + C_2 e^{3x}$. Thus we choose for the particular integral

$$y_p(x) = e^x\,(A \cos 2x + B \sin 2x) + e^{2x}\,[(Cx + D)\cos x + (Ex + F)\sin x] \ .$$

T 4.5 The auxiliary equation is $\lambda^2 - 4\lambda + 4 = 0$ with the double root $\lambda_{1,2} = 2$. The complementary function is $y_c(x) = C_1 e^{2x} + C_2 x e^{2x}$ and the particular integral is chosen to be

$$
\begin{aligned}
y_p(x) \;=\;& \left(Ax^2 + Bx + C\right) + x^2\,e^{2x}\,(Dx + E) \\
&+ (Fx + G)\sin 2x + (Hx + J)\cos 2x \ .
\end{aligned}
$$

T 4.6 The auxiliary equation is $\lambda^2 + 3\lambda + 2 = 0$ with roots $\lambda_1 = -1$, $\lambda_2 = -2$. The complementary function is $y_c(x) = C_1 e^{-x} + C_2 e^{-2x}$ and the particular integral is

$$
\begin{aligned}
y_p(x) \;=\;& \left(Ax^2 + Bx + C\right) e^x \sin 2x + \left(Dx^2 + Ex + F\right) e^x \cos 2x \\
&+ (G \sin x + H \cos x)\,e^{-x} + Je^x \ .
\end{aligned}
$$

T 4.7 Solving the homogeneous equation first, $y = a\,e^{\alpha t}$ leads to the auxiliary equation $\alpha^2 + \lambda^2 = 0$ with the roots $\alpha_{1,2} = \pm i\lambda$. Therefore, the complementary function is

$$y_c(x) = A \cos \lambda x + B \sin \lambda x \ .$$

Let

$$y_p(x) = \sum_{m=1}^{N} y_{pm}(x) \qquad \longrightarrow \qquad y_{pm}'' + \lambda^2 y_{pm} = a_m \sin m\pi x \ .$$

Under the assumption $\lambda \neq m\pi$ we can then choose

$$y_{pm}(x) = \alpha_m \cos m\pi x + \beta_m \sin m\pi x.$$

The usual procedure of substitution and comparison of coefficients yields $\alpha_m(\lambda^2 - m^2\pi^2) = 0$ and $\beta_m(\lambda^2 - m^2\pi^2) = \alpha_m$. The general solution is therefore

$$y(x) = A\cos \lambda x + B\sin \lambda x + \sum_{m=1}^{N} \frac{\alpha_m}{\lambda^2 - m^2\pi^2} \sin m\pi x.$$

(Expressions of that type are important in the theory of Fourier series.)

T 4.8   Rewrite the differential equation as

$$y'' + \frac{y}{4} = \frac{1}{2\cos(x/2)}.$$

The complementary function is

$$y_c(x) = A\cos(x/2) + B\sin(x/2).$$

With $y_1(x) = \cos(x/2)$ and $y_2(x) = \sin(x/2)$ we get

$$W(y_1, y_2) = \cos(x/2)\,\frac{1}{2}\cos(x/2) - \left(-\frac{1}{2}\right)\sin(x/2)\sin(x/2) = \frac{1}{2}$$

and $g(x) = 1/[2\cos(x/2)]$. Hence

$$u_1(x) \;=\; -\int \frac{\sin(x/2)}{(1/2)}\,\frac{1}{2\cos(x/2)}\,dx = -\int \frac{\sin(x/2)}{\cos(x/2)}\,dx = 2\ln[\cos(x/2)],$$

$$u_2(x) \;=\; \int \frac{\cos(x/2)}{(1/2)}\,\frac{1}{2\cos(x/2)}\,dx = \int dx = x.$$

The general solution is therefore

$$
\begin{aligned}
y(x) &= y_c(x) + u_1(x)y_1(x) + u_2(x)y_2(x) \\
&= A\cos(x/2) + B\sin(x/2) + 2\cos(x/2)\ln[\cos(x/2)] + x\sin(x/2).
\end{aligned}
$$

T 4.9   The auxiliary equation $\lambda^2 - 2\lambda + 1 = 0$ leads to the complementary function

$$y_c(x) = A\,e^x + Bx\,e^x.$$

With $g(x) = e^x/(1+x^2)$ and $y_1 = e^x$, $y_2 = xe^x$ it may be shown that $W(y_1, y_2) = y_1 y_2' - y_1' y_2 = e^{2x}$. Therefore,

$$u_1(x) = -\int \frac{xe^x \left[e^x/(1+x^2)\right]}{e^{2x}} dx = -\int \frac{x}{1+x^2} dx = -\frac{1}{2}\ln(1+x^2),$$

$$u_2(x) = \int \frac{e^x \left[e^x/(1+x^2)\right]}{e^{2x}} dx = \int \frac{1}{1+x^2} dx = \tan^{-1} x.$$

This leads to the general solution

$$y(x) = y_c(x) + u_1(x)y_1(x) + u_2(x)y_2(x)$$
$$= Ae^x + Bxe^x - \frac{1}{2}e^x \ln(1+x^2) + xe^x \tan^{-1} x.$$

T 4.10   Substitution of the given solutions and their first and second derivatives shows that they both fulfil the homogeneous equation. We rewrite the differential equation as

$$y'' + \frac{x}{1-x}y' - \frac{1}{1-x}y = 2(1-x)e^{-x}.$$

Therefore, $g(x) = 2(1-x)e^{-x}$ and the Wronskian determinant is $W(y_1, y_2) = y_1 y_2' - y_1' y_2 = (1-x)e^x$. Thus

$$u_1(x) = -\int \frac{x\,2(1-x)e^{-x}}{(1-x)e^x} dx = -2\int xe^{-2x} dx = \frac{1}{2}(2x+1)e^{-2x},$$

$$u_2(x) = \int \frac{e^x\,2(1-x)e^{-x}}{e^x(1-x)} dx = 2\int e^{-x} dx = -2e^{-x}.$$

Consequently, a particular integral is

$$y_p(x) = u_1(x)y_1(x) + u_2(x)y_2(x) = -\frac{1}{2}(2x-1)e^{-x}.$$

T 4.11   The verification procedure works as in the previous example. We see that $g(x) = (12x^{3/2}\sin x)/(4x^2) = 3x^{-1/2}\sin x$ and obtain the Wronskian $W(y_1, y_2) = -x^{-1}$. Therefore,

$$u_1(x) = -\int \frac{x^{-1/2}\cos x\,3x^{-1/2}\sin x}{(-x^{-1})} dx = -\frac{3}{2}\cos^2 x,$$

$$u_2(x) = \int \frac{x^{-1/2}\sin x\,3x^{-1/2}\sin x}{(-x^{-1})} dx = -\frac{3}{2}(x - \sin x \cos x).$$

A particular integral is therefore

$$y_p(x) = u_1(x)y_1(x) + u_2(x)y_2(x) = -\frac{3}{2}x^{1/2}\cos x \, .$$

T 4.12   It is evident that $y = e^{-2x}$ is a solution of the homogeneous equation and therefore it makes sense to write $y = ve^{-2x}$. Thus v satisfies

$$xe^{-2x}\left(\frac{d^2v}{dx^2} - 4\frac{dv}{dx} + 4v\right) + e^{-2x}\left(\frac{dv}{dx} - 2v\right) + (2 - 4x)ve^{-2x} = 8x(1-x)\, .$$

Let $z = dv/dx$ then, after some algebra, z is found to satisfy

$$\frac{dz}{dx} + \left(\frac{1}{x} - 4\right)z = 8(1-x)e^{2x}\, .$$

This is a linear equation with integrating factor xe^{-4x}. Therefore

$$\frac{d}{dx}(x\,e^{-4x}\,z) = 8x(1-x)\,e^{-2x} \qquad \longrightarrow \qquad x\,e^{-4x}\,z = 4x^2\,e^{-2x} + A\, .$$

Thus

$$\frac{dv}{dx} = 4xe^{2x} + A\frac{e^{4x}}{x} \qquad \longrightarrow \qquad v = (2x-1)\,e^{2x} + A\int^x \frac{e^{4s}}{s}\,ds + B\, .$$

The general solution to the original differential equation is therefore

$$y = A\,e^{-2x}\int^x \frac{e^{4s}}{s}\,ds + B\,e^{-2x} + 2x - 1\, .$$

T 4.13   The equation is of Euler's type with solution $y = x^n$. Then

$$x^2(n^2 - n)x^{n-2} + 3nx^{n-1} + x^n = (n^2 + 2n + 1)x^n = 0$$

so that $n = -1$ is a double root. Now let $y = v/x$, then v satisfies

$$x^2\left(\frac{1}{x}\frac{d^2v}{dx^2} - \frac{2}{x^2}\frac{dv}{dx} + \frac{2}{x^3}v\right) + 3x\left(\frac{1}{x}\frac{dv}{dx} - \frac{1}{x^2}v\right) + \frac{v}{x} = \ln x\, .$$

Let $z = dv/dx$ then, after some algebra, z is found to satisfy

$$\frac{dz}{dx} + \frac{z}{x} = \frac{\ln x}{x}\, .$$

This is a linear equation with integrating factor x and whose solution is found by the calculation

$$\frac{d}{dx}(xz) = \ln x \qquad \longrightarrow \qquad \frac{dv}{dx} = z = \ln x - 1 + \frac{A}{x}.$$

This equation for v integrates to

$$v = x \ln x - 2x + A \ln x + B$$

and thus the original equation has general solution

$$y = -2 + A\frac{\ln x}{x} + \frac{B}{x} + \ln x.$$

T 4.14 Substitution of $y = x^\lambda$ and its first two derivatives into the differential equation gives

$$\left(\lambda^2 - \lambda + 4\lambda + 2\right) x^\lambda = 0.$$

Therefore $\lambda^2 + 3\lambda + 2 = 0$ with the roots $\lambda_1 = -1$ and $\lambda_2 = -2$. The general solution is then

$$y(x) = \frac{A}{x} + \frac{B}{x^2} \qquad \longrightarrow \qquad y'(x) = -\frac{A}{x^2} - \frac{2B}{x^3}.$$

We have

$$y(a) = a \quad \longrightarrow \quad \frac{A}{a} + \frac{B}{a^2} = a, \qquad y'(a) = -1 \quad \longrightarrow \quad -\frac{A}{a^2} - \frac{2B}{a^3} = -1$$

which leads to $A = a^2$ and $B = 0$. The particular solution is therefore

$$y(x) = \frac{a^2}{x}.$$

T 4.15 Once again we use $y = x^\lambda$ in the homogeneous equation and get $\lambda^2 - 2\lambda - 3 = 0$ with roots $\lambda_1 = 3$ and $\lambda_2 = -1$ and a general solution of

$$y(x) = Ax^3 + \frac{B}{x} + y_p(x).$$

For the particular integral y_p, we choose $y_p = \alpha x^2$. Upon substitution of y_p into the inhomogeneous equation and comparison of coefficients, we find $\alpha = -1/3$. The initial conditions yield

$$y(1) = 0 \quad \longrightarrow \quad A + B - \frac{1}{3} = 0, \qquad y'(1) = 0 \quad \longrightarrow \quad 3A - B - \frac{2}{3} = 0.$$

This equation for ϕ simplifies to

$$x\phi'' + 2\phi' - 2(1+x)\phi' - 2\phi + 2(1+x)\phi = x(\phi'' - 2\phi' + 2\phi) = 0.$$

Therefore, any solution of $\phi'' - 2\phi' + 2\phi = 0$ is an integrating factor of the original differential equation. The function $\phi(x) = e^x \sin x$ (the roots of the auxiliary equation are $1 \pm i$) is one such solution, and in this instance, the original differential equation may now be rewritten

$$\frac{d}{dx}\left[x\,e^x\,\sin x\,\frac{dy}{dx} + e^x\left[(1+x)\sin x - x\cos x\right]y\right] = 0.$$

The last equation can be integrated and we obtain

$$x\,e^x\,\sin x\,\frac{dy}{dx} + e^x\left[(1+x)\sin x - x\cos x\right]y = A.$$

This equation is now re-expressed as the linear first order equation

$$\frac{dy}{dx} + \left(\frac{1}{x} + 1 - \frac{\cos x}{\sin x}\right)y = \frac{A\,e^{-x}}{x\,\sin x}$$

with integrating factor $x\,e^x/\sin x$. Therefore,

$$\frac{d}{dx}\left(\frac{x\,e^x\,y}{\sin x}\right) = \frac{A}{\sin^2 x} \qquad \longrightarrow \qquad \frac{x\,e^x\,y}{\sin x} = -A\,\frac{\cos x}{\sin x} + B.$$

The general solution is now

$$y(x) = \frac{e^{-x}}{x}\left(-A\cos x + B\sin x\right).$$

If y has a finite limit as $x \to 0$ then $A = 0$ and the value of this limit is determined by L'Hôpital's rule as unity if $B = 1$. Therefore the required particular solution is

$$y(x) = \frac{e^{-x}\sin x}{x}.$$

T 4.20   The given equation is exact because $(x^2)'' - (4x+4x^2)' + 8x + 2 = 0$ and may be re-expressed as

$$\frac{d}{dx}\left[x^2\frac{dy}{dx} + (4x^2 + 2x)y\right] = 0.$$

The integrated equation is now

$$x^2\frac{dy}{dx} + (4x^2 + 2x)y = C \qquad \longrightarrow \qquad \frac{dy}{dx} + \left(4 + \frac{2}{x}\right)y = \frac{C}{x^2}.$$

This equation has integrating factor $x^2 e^{4x}$ and therefore

$$\frac{d}{dx}\left(x^2 e^{4x} y\right) = Ce^{4x} \qquad \longrightarrow \qquad x^2 e^{4x} y = Ae^{4x} + B\,.$$

Thus the general solution is

$$y(x) = \frac{A}{x^2} + B\,\frac{e^{-4x}}{x^2}\,.$$

The initial conditions yield

$$y(1) = 0 \;\longrightarrow\; A + Be^{-4} = 0, \qquad y'(1) = 1 \;\longrightarrow\; -2A - 6Be^{-4} = 1\,.$$

Hence $A = 1/4$ and so $B = -e^4/4$. The particular solution is

$$y = \frac{1 - e^{4(1-x)}}{4x^2}\,.$$

T 4.21   We have

$$u = e^{-\int py\,dx} \qquad \longrightarrow \qquad u' = -py\, e^{-\int py\,dx}$$

and thus

$$y = -\frac{u'}{pu} \qquad \longrightarrow \qquad y' = -\frac{u''}{pu} + \frac{u'^2}{pu^2} + \frac{p'u'}{p^2 u}\,.$$

Upon substitution, the result is proven. In the example we set $p = -e^x$, $q = -1$ and $r = 4e^{-x}$. This leads to $u'' - 4u = 0$ with the general solution $u = Ae^{2x} + Be^{-2x}$. Therefore

$$y(x) = -\frac{u'}{pu} = 2\,e^{-x}\,\frac{Ae^{2x} - Be^{-2x}}{Ae^{2x} + Be^{-2x}} = 2\,e^{-x}\,\frac{Ce^{4x} - 1}{Ce^{4x} + 1}\,.$$

T 4.22   Since $p = k\rho^2$ then $p' = 2k\rho\rho'$. Momentum balance $p' = -g\rho$ now indicates that $2k\rho' = -g$. Multiplying this equation by r^2 and differentiating with respect to r yields

$$2k(r^2\rho')' = -(r^2 g)' = -4\pi G r^2 \rho\,.$$

Therefore

$$r^2 \rho'' + 2r\rho' + \alpha^2 r^2 \rho = 0 \qquad \longrightarrow \qquad r\rho'' + 2\rho' + \alpha^2 r\rho = 0$$

with $\alpha^2 = 2\pi G/k$. Suppose that $\phi(r)$ is an integrating factor of this equation, then we must have

$$(r\phi)'' - 2\phi' + \alpha^2 r\phi = 0 \qquad \longrightarrow \qquad \phi'' + \alpha^2 \phi = 0\,.$$

Take $\phi(r) = \sin \alpha r$, then the original differential equation for ρ can be re-expressed as

$$\frac{d}{dr}\left[r \sin \alpha r \, \rho' + (\sin \alpha r - \alpha r \cos \alpha r)\rho\right] = 0$$

$$\longrightarrow \quad r \sin \alpha r \, \rho' + (\sin \alpha r - \alpha r \cos \alpha r)\rho = A$$

$$\longrightarrow \quad \rho' + \left(\frac{1}{r} - \alpha \cot \alpha r\right)\rho = \frac{A}{r \sin \alpha r}.$$

The integrating factor of this equation is $r/\sin \alpha r$. Hence

$$\frac{d}{dr}\left(\frac{r\rho}{\sin \alpha r}\right) = \frac{A}{\sin^2 \alpha r} \quad \longrightarrow \quad \rho(r) = -A\frac{\cos \alpha r}{\alpha r} + B\frac{\sin \alpha r}{r}.$$

Since $\rho(r)$ is a density, it must be non–negative and finite everywhere within the star. In particular $\rho(r)$ must be finite as $r \to 0$. This latter condition requires that $A = 0$ by L'Hôpital's rule. Therefore the final density is

$$\rho(r) = B\frac{\sin \alpha r}{r} = \rho(0)\frac{\sin \alpha r}{\alpha r}$$

where $\rho(0)$ is the density at the core of the star. The density on its exterior boundary is $\rho(0)(\sin \alpha a)/(\alpha a)$. This density is non–negative only provided $\alpha a \le \pi$. This stellar model therefore predicts stars of maximum radius π/α.

TUTORIAL EXAMPLES 5

T 5.1   The auxiliary equation $\lambda^2 + 2\alpha\lambda + \omega_0^2 = 0$ has roots $\lambda_{1,2} = -\alpha \pm i\Omega$ where $\Omega = \sqrt{\omega_0^2 - \alpha^2}$. The general solution is

$$g(t) = e^{-\alpha t}\left(A \sin \Omega t + B \cos \Omega t\right).$$

It follows from $g(0) = 0$ that $B = 0$, whereas $\dot{g}(0) = \beta$ requires $A = \beta/\Omega$. Thus the particular solution is

$$g(t) = \beta e^{-\alpha t}\frac{\sin \Omega t}{\Omega}.$$

Note that the equation describes the *deviation* of glucose level from the fasted level and therefore the values that are given for the patient are $g(1) = 0.3$, $g(2) = -0.15$ and $g(3) = 0.05$. Substitution into the particular solution gives

$$\beta e^{-\alpha}\frac{\sin \Omega}{\Omega} = 0.3, \quad \beta e^{-2\alpha}\frac{\sin 2\Omega}{\Omega} = -0.15, \quad \beta e^{-3\alpha}\frac{\sin 3\Omega}{\Omega} = 0.05.$$

This is a system of three (non–linear) algebraic equations for the three un-known parameters. Multiplication of the first with the third expression and division by the square of the second one leads to

$$\frac{\sin \Omega \, \sin 3\Omega}{\sin^2 2\Omega} = \frac{2}{3}.$$

Letting $s = \sin \Omega \neq 0$, we obtain

$$\frac{s \, (3s - 4s^3)}{4s^2 \, (1 - s^2)} = \frac{2}{3} \qquad \longrightarrow \qquad 4s^2 = 1 \qquad \longrightarrow \qquad s_{1,2} = \pm \frac{1}{2}.$$

Consequently, there are a total of four solutions for Ω: $\pi/6$, $5\pi/6$, $-\pi/6$, $-5\pi/6$. The latter two, being negative, are not relevant since Ω is positive by definition. If, further, $\Omega = \pi/6$, then $\sin 2\Omega = \sin(\pi/3) > 0$ which is not possible (see the second of the above parameter relations). Hence, the only remaining solution is $\Omega = 5\pi/6$. We can then use the first two parameter relations above to calculate α and β and get

$$\alpha = \frac{1}{2} \ln 12, \qquad \beta = \sqrt{3} \, \pi, \qquad \Omega = 5\pi/6.$$

Therefore, $\omega_0^2 = \Omega^2 + \alpha^2 \approx 8.40$, $\omega_0 \approx 2.90$ and the period of the un-damped motion is $2\pi/\omega_0 \approx 2.14$ hours, which makes the patient strongly non–diabetic.

T 5.2 The differential equation governing the behaviour of this system is

$$\ddot{y} + 2\alpha\dot{y} + \omega_0^2 y = 0$$

with suitable initial conditions. The solution is oscillatory, so $\omega_0^2 > \alpha^2$. The damping is also light and so we have the general solution

$$y(t) = A \, e^{-\alpha t} \, \sin(\Omega t + \beta)$$

with $\Omega = \sqrt{\omega_0^2 - \alpha^2}$. The period is $T = 2\pi/\Omega$ and the number of oscillations to half amplitude is N where

$$\frac{1}{2} = e^{-\alpha 2\pi N/\Omega} \qquad \longrightarrow \qquad N = \frac{\Omega \ln 2}{2\pi\alpha}.$$

Required is a discussion of the behaviour of T and N as α and ω_0^2 (which is directly proportional to the restoring force) are changed. First, we write

$$\ln T = \ln 2\pi - \frac{1}{2} \ln(\omega_0^2 - \alpha^2).$$

Therefore

$$\frac{1}{T}\frac{\partial T}{\partial \alpha} = \frac{\alpha}{\omega_0^2 - \alpha^2}, \qquad \frac{1}{T}\frac{\partial T}{\partial \omega_0} = -\frac{\omega_0}{\omega_0^2 - \alpha^2}.$$

These expression now permit an analysis of the period. The period is locally an increasing function of α and a decreasing function of restoring effects. Of course, if the damping is increased too much, the solution is no longer oscillatory. A stronger restoring influence makes the period shorter.

To analyse the vibration count we derive

$$\ln N = \ln\left[(\ln 2)/2\pi\right] + \frac{1}{2}\ln\left(\omega_0^2 - \alpha^2\right) - \ln \alpha$$

and therefore

$$\frac{1}{N}\frac{\partial N}{\partial \alpha} = -\frac{\alpha}{\omega_0^2 - \alpha^2} - \frac{1}{\alpha} = -\frac{\omega_0^2}{\alpha\left(\omega_0^2 - \alpha^2\right)} < 0,$$

$$\frac{1}{N}\frac{\partial N}{\partial \omega_0} = \frac{\omega_0}{\omega_0^2 - \alpha^2} > 0.$$

The number of oscillations to half amplitude decreases as α increases and increases as the restoring effects increase. Again, these are local results.

Finally, for the extraction of the differential equation we set $T = 1$ and $N = 40$. We can calculate $\alpha \approx 0.0173$ and $\omega_0 \approx \Omega \approx 6.283$. From the above expressions the differential equation $\ddot{y} + 0.0346\,\dot{y} + 39.479\,y = 0$ is obtained.

TUTORIAL EXAMPLES 6

T 6.1

$$L\left[e^{-t-1/2} : t \to s\right] = \int_0^\infty e^{-st}\,e^{-t-1/2}\,dt$$

$$= e^{-1/2}\int_0^\infty e^{-t(s+1)}\,dt = e^{-1/2}\left[\frac{e^{-t(s+1)}}{-(s+1)}\right]_0^\infty = \frac{e^{-1/2}}{s+1}$$

with the requirement $s > -1$ for the integral to be finite.

T 6.2   With $\cos(at + b) = \cos at \, \cos b - \sin at \, \sin b$ we have

$$L\left[\cos(at + b) : t \to s\right]$$

$$= \cos b\, L\left[\cos at : t \to s\right] - \sin b\, L\left[\sin at : t \to s\right]$$

$$= \cos b \int_0^\infty e^{-st}\cos at\, dt - \sin b \int_0^\infty e^{-st}\sin at\, dt$$

$$= \cos b\,\frac{s}{s^2 + a^2} - \sin b\,\frac{a}{s^2 + a^2}$$

whereby the integrals can be evaluated either by standard methods or use can be made of the Laplace transform Table 6.1. Note that the integrals exist only if $s > 0$.

T 6.3   With the substitution $u = st$ we find

$$L\left[t^p : t \to s\right] = \int_0^\infty e^{-st}\, t^p\, dt = \int_0^\infty e^{-u} \frac{u^p}{s^p} \frac{1}{s}\, du$$

$$= \frac{1}{s^{p+1}} \int_0^\infty e^{-u}\, u^p\, du = \frac{1}{s^{p+1}}\, \Gamma(p+1)$$

by virtue of the definition of the *Gamma function* ($p > -1$, $s > 0$). For an integer n we have

$$\Gamma(n+1) = n\Gamma(n) = n(n-1)\Gamma(n-2) = \cdots$$

$$= n(n-1)(n-2) \times \cdots \times 3 \times 2 \times 1 \times \Gamma(1) = n!$$

so that $L\left[t^n : t \to s\right] = n!/s^{n+1}$. Next, for $p = -1/2$:

$$L\left[t^{-1/2} : t \to s\right] = s^{-1/2}\, \Gamma\left(\frac{1}{2}\right) = \left(\frac{\pi}{s}\right)^{1/2}.$$

For $p = 1/2$:

$$L\left[t^{1/2} : t \to s\right] = s^{-3/2}\, \Gamma\left(\frac{3}{2}\right) = s^{-3/2} \frac{1}{2}\, \Gamma\left(\frac{1}{2}\right) = \frac{\pi^{1/2}}{2s^{3/2}}.$$

And for $p = 5/2$:

$$L\left[t^{5/2} : t \to s\right] = s^{-7/2}\, \Gamma\left(\frac{7}{2}\right) = s^{-7/2} \frac{5}{2} \frac{3}{2} \frac{1}{2}\, \Gamma\left(\frac{1}{2}\right) = \frac{15\pi^{1/2}}{8s^{7/2}}.$$

T 6.4   We have

$$\bar{f}(s) = \frac{s-2}{s^2-2} = \frac{s-2}{(s-\sqrt{2})(s+\sqrt{2})}$$

$$= \frac{1-\sqrt{2}}{2} \frac{1}{s-\sqrt{2}} + \frac{1+\sqrt{2}}{2} \frac{1}{s+\sqrt{2}}$$

by using a partial fraction decomposition. Since $L\left[e^{\alpha t} : t \to s\right] = 1/(s-\alpha)$, we obtain

$$f(t) = L^{-1}\left[\bar{f}(s) : s \to t\right] = L^{-1}\left[\frac{s-2}{s^2-2} : s \to t\right]$$

$$= \frac{1-\sqrt{2}}{2} L^{-1}\left[\frac{1}{s-\sqrt{2}} : s \to t\right] + \frac{1+\sqrt{2}}{2} L^{-1}\left[\frac{1}{s+\sqrt{2}} : s \to t\right]$$

$$= \frac{1-\sqrt{2}}{2} e^{\sqrt{2}t} + \frac{1+\sqrt{2}}{2} e^{-\sqrt{2}t}$$

$$= \cosh\sqrt{2}\,t - \sqrt{2}\,\sinh\sqrt{2}\,t.$$

with the help of the Heaviside function. Thus

$$L[f(t):t \to s] = \int_0^\infty [1 + 2H(t-1) + 2H(t-7)]\, e^{-st}\, dt$$

$$= \int_0^\infty e^{-st}\, dt + 2\int_1^\infty e^{-st}\, dt + 2\int_7^\infty e^{-st}\, dt$$

$$= \frac{1}{s} + \frac{2e^{-s}}{s} + \frac{2e^{-7s}}{s} = \frac{1}{s}\left(1 + 2e^{-s} + 2e^{-7s}\right),$$

valid for $s > 0$.

T 6.11   We can write $g(t) = 1 - H(t - \pi)$ and obtain

$$\left[s^2\bar{y} - sy(0) - \dot{y}(0)\right] + \left[s\bar{y} - sy(0)\right] + \frac{5}{4}\bar{y} = \frac{1}{s} - \frac{e^{-\pi s}}{s}.$$

Then, after substitution of the initial values and collection of terms we get

$$\bar{y} = \left(1 - e^{-\pi s}\right)K(s)$$

where

$$K(s) = \frac{1}{s(s^2 + s + 5/4)} = \frac{4}{5s} - \frac{4}{5}\frac{s+1}{(s+1/2)^2 + 1}.$$

We note that

$$y(t) = L^{-1}\left[\left(1 - e^{-\pi s}\right)K(s):s \to t\right]$$

$$= L^{-1}[K(s):s \to t] - L^{-1}\left[e^{-\pi s}K(s):s \to t\right]$$

$$= k(t) - k(t - \pi)H(t - \pi)$$

where

$$k(t) = L^{-1}[K(s):s \to t]$$

$$= L^{-1}\left[\frac{4}{5s}:s \to t\right] - L^{-1}\left[\frac{4}{5}\frac{s+1/2+1/2}{(s+1/2)^2 + 1}:s \to t\right]$$

$$= \frac{4}{5}L^{-1}\left[\frac{1}{s}:s \to t\right] - \frac{4}{5}L^{-1}\left[\frac{s+1/2}{(s+1/2)^2 + 1}:s \to t\right]$$

$$\qquad - \frac{2}{5}L^{-1}\left[\frac{1}{(s+1/2)^2 + 1}:s \to t\right]$$

$$= \frac{4}{5} - \frac{4}{5}e^{-t/2}\cos t - \frac{2}{5}e^{-t/2}\sin t,$$

using the first shifting property. Therefore we find the solution as

$$y(t) = \frac{2}{5}\left[2 - e^{-t/2}\left(2\cos t + \sin t\right)\right]$$

$$\qquad - \frac{2H(t-\pi)}{5}\left[2 + e^{-(t-\pi)/2}\left(2\cos t + \sin t\right)\right].$$

If so desired, the solution can be split into its two constituent parts:

$$y(t) = \begin{cases} \frac{2}{5}\left[2 - e^{-t/2}\left(2\cos t + \sin t\right)\right] & 0 \le t < \pi \\ -\frac{2}{5}e^{-t/2}\left(2\cos t + \sin t\right)\left(1 + e^{\pi/2}\right) & t \ge \pi, \end{cases}$$

as illustrated in Figure B.2.

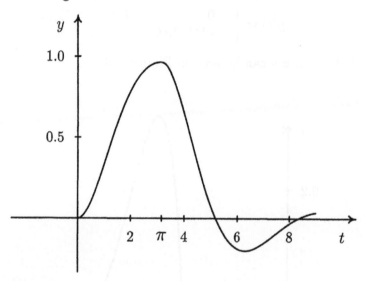

Figure B.2: Solution of Tutorial example T 6.11, y versus t.

T 6.12 By using the definition of the Laplace transform we obtain

$$L\left[f(t-a)H(t-a) : t \to s\right] = \int_0^\infty f(t-a)H(t-a)e^{-st}\,dt$$

$$= \int_a^\infty f(t-a)e^{-st}\,dt = \int_0^\infty f(u)e^{-s(a+u)}\,du$$

$$= e^{-as}\int_0^\infty f(u)e^{-su}\,du = e^{-as}\overline{f}(s)$$

where in going from the first to the second term on the second line in this calculation, the substitution $t - a = u$, $dt = du$ was used.

T 6.13 Application of the Laplace transform yields

$$\left(s^2 + 2s + 2\right)L\left[y : t \to s\right] = e^{-\pi s}.$$

With

$$k(t) = L^{-1}\left[\frac{1}{(s+1)^2+1} : s \to t\right] = e^{-t}L^{-1}\left[\frac{1}{s^2+1} : s \to t\right] = e^{-t}\sin t$$

we have

$$y(t) = L^{-1}\left[\frac{e^{-\pi s}}{(s+1)^2 + 1} : s \to t\right] = H(t - \pi)\, k(t - \pi)$$

$$= H(t - \pi)e^{\pi - t}\, \sin(t - \pi).$$

The solution can therefore be written as

$$y(t) = \begin{cases} 0 & t < \pi \\ -e^{\pi - t}\, \sin t & t \geq \pi. \end{cases}$$

A sketch of the solution can be seen in Figure B.3.

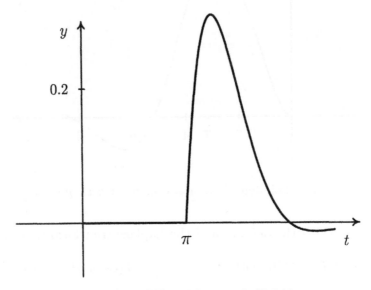

Figure B.3: Solution of Tutorial example T 6.13, y versus t.

T 6.14   For the simple circuit consisting of an inductor and a capacitor and the scenario as described, we obtain the initial value problem

$$\ddot{Q}(t) + \frac{1}{LC}\, Q(t) = 0, \qquad Q(0) = Q_0, \quad \dot{Q}(0) = 0.$$

Application of the Laplace transform to this problem yields

$$s^2\, \overline{Q}(s) - s\, Q_0 + \gamma^2\, \overline{Q}(s) = 0$$

where $\overline{Q}(s) = L[Q(t) : t \to s]$ and $\gamma = 1/\sqrt{LC}$. Therefore,

$$\overline{Q}(s) = \frac{s\, Q_0}{s^2 + \gamma^2} \qquad \longrightarrow \qquad Q(t) = Q_0\, \cos \gamma t.$$

The solution represents an oscillation with frequency γ (energy is stored alternatively in the capacitor and the inductor but the total amount is conserved.)

In the more general situation of the LCR circuit, the corresponding initial value problem is

$$\ddot{Q}(t) + \frac{R}{L}\dot{Q}(t) + \frac{1}{LC}Q(t) = 0, \qquad Q(0) = Q_0, \quad \dot{Q}(0) = 0.$$

Here the application of the Laplace transform method gives

$$\overline{Q}(s) = \frac{s\,Q_0 + (R/L)Q_0}{s^2 + (R/L)s + \gamma^2}.$$

Denote the two roots of the quadratic polynomial in the denominator by

$$s_{1,2} = -\frac{R}{2L} \pm i\delta, \qquad \delta^2 = \frac{1}{LC} - \frac{R^2}{4L^2} > 0,$$

then a partial fraction expansion yields

$$\begin{aligned}
\overline{Q}(s) &= \frac{Q_0}{s_1 - s_2}\left(\frac{s_1}{s - s_1} - \frac{s_2}{s - s_2} + \frac{R}{L}\frac{1}{s - s_1} - \frac{R}{L}\frac{1}{s - s_2}\right) \\
&= \frac{Q_0}{s_1 - s_2}\left(\frac{s_1 + R/L}{s - s_1} - \frac{s_2 + R/L}{s - s_2}\right).
\end{aligned}$$

The solution for $Q(t)$ is therefore

$$Q(t) = \frac{Q_0}{s_1 - s_2}\left[\left(s_1 + \frac{R}{L}\right)e^{s_1 t} - \left(s_2 + \frac{R}{L}\right)e^{s_2 t}\right].$$

Substituting the values for $s_{1,2}$ and rearranging terms, one finds that

$$Q(t) = Q_0\, e^{-(R/2L)t}\left(\cos\delta t + \frac{R}{2L\delta}\sin\delta t\right).$$

The effect of the resistor on the circuit is to provide a *damping* mechanism. As current flows through the resistor, energy is transformed into heat and is therefore lost to the system. The amplitude of the oscillations will decrease exponentially. If $R^2 \ll 4L/C$ then $\delta \approx \gamma$, so that the frequency of the damped oscillator is very close to that of the undamped oscillator. Of course, in the limit $R \to 0$, the solution for the LCR circuit reduces to that of the LC circuit given earlier.

T 6.15 The current $I(t)$ in the circuit is governed by the differential equation

$$L\dot{I} + RI = U_{\text{square}}(t), \qquad I(0) = 0.$$

The Laplace transform $\overline{I}(s) = L[I(t) : t \to s]$ therefore satisfies the algebraic equation

$$(Ls + R)\overline{I} = \frac{V}{s} \frac{1}{1 + e^{-sT}}.$$

Define $a = R/L$ then $\overline{I}$ is given by

$$\overline{I} = \frac{V}{L} \frac{1}{s(s+a)} \frac{1}{(1 + e^{-sT})} = \frac{V}{L} \sum_{k=0}^{\infty} (-1)^k \frac{e^{-skT}}{s(s+a)}.$$

By recognizing that

$$\frac{a}{s(s+a)} = \frac{1}{s} - \frac{1}{(s+a)},$$

the Laplace transform $\overline{I}$ may be inverted to obtain

$$I(t) = \frac{V}{R} \sum_{k=0}^{\infty} (-1)^k \left[1 - e^{-a(t-kT)} \right] H(t - kT).$$

In this solution, H is the Heaviside function.

TUTORIAL EXAMPLES 7

T 7.1   We set $z = y''$ and the differential equation becomes $xz' - z = 0$, which is a linear equation with general solution $z = Cx$. Integration gives, first, $y' = Dx^2 + F$, and second, $y = Ex^3 + Fx + G$. We can thus identify $y_1(x) = 1$, $y_2(x) = x$, $y_3(x) = x^3$. The Wronskian is

$$W = \begin{vmatrix} y_1 & y_2 & y_3 \\ y_1' & y_2' & y_3' \\ y_1'' & y_2'' & y_3'' \end{vmatrix} = \begin{vmatrix} 1 & x & x^3 \\ 0 & 1 & 3x^2 \\ 0 & 0 & 6x \end{vmatrix} = \begin{vmatrix} 1 & 3x^2 \\ 0 & 6x \end{vmatrix} = 6x.$$

T 7.2   The differential equation is of Euler's type. We substitute $y = x^\lambda$ and obtain $\lambda^3 - 2\lambda^2 - \lambda + 2 = 0$ with roots $\lambda_1 = 1$, $\lambda_2 = 2$, $\lambda_3 = -1$. We can thus identify $y_1(x) = x$, $y_2(x) = x^2$, $y_3(x) = 1/x$. The Wronskian is

$$W = \begin{vmatrix} x & x^2 & x^{-1} \\ 1 & 2x & -x^{-2} \\ 0 & 2 & 2x^{-3} \end{vmatrix} = x \begin{vmatrix} 2x & x^{-2} \\ 2 & 2x^{-3} \end{vmatrix} - \begin{vmatrix} x^2 & x^{-1} \\ 2 & 2x^{-3} \end{vmatrix} = \frac{6}{x}.$$

T 7.3   The characteristic equation is $\lambda^4 + 1 = 0$. Its four roots, written in polar form, are

$$\lambda_{1,2,3,4} = e^{i(\pi/4 + m\pi/2)}, \qquad m = 0, 1, 2, 3$$

or

$$\lambda_1 = \frac{1+i}{\sqrt{2}}, \quad \lambda_2 = \frac{-1+i}{\sqrt{2}}, \quad \lambda_3 = \frac{-1-i}{\sqrt{2}}, \quad \lambda_4 = \frac{1-i}{\sqrt{2}}.$$

The general solution is therefore

$$y(x) = Ae^{\lambda_1 x} + Be^{\lambda_2 x} + Ce^{\lambda_3 x} + De^{\lambda_4 x},$$

or, by splitting into real and imaginary parts

$$y(x) = e^{x/\sqrt{2}} \left[a\cos(x/\sqrt{2}) + b\sin(x/\sqrt{2}) \right]$$
$$+ e^{-x/\sqrt{2}} \left[c\cos(x/\sqrt{2}) + d\sin(x/\sqrt{2}) \right].$$

T 7.4   The auxiliary equation is

$$\lambda^4 - 8\lambda = \lambda(\lambda^3 - 8) = \lambda(\lambda - 2)(\lambda^2 + 2\lambda + 4) = 0$$

with roots $\lambda_1 = 0$, $\lambda_2 = 2$, $\lambda_{3,4} = -1 \pm i\sqrt{3}$. The general solution is therefore

$$y(x) = A + Be^{2x} + Ce^{(-1+i\sqrt{3})x} + De^{(-1-i\sqrt{3})x}$$

or

$$y(x) = A + Be^{2x} + e^{-x}\left(c\cos\sqrt{3}x + d\sin\sqrt{3}x\right).$$

T 7.5   The auxiliary equation is $\lambda^4 - 2\lambda^2 + 1 = (\lambda^2 - 1)^2 = 0$ and has (repeated) roots $\lambda_{1,2,3,4} = 1, 1, -1, -1$. Therefore

$$y(x) = Ae^x + Be^{-x} + Cxe^x + Dxe^{-x}$$

or

$$y(x) = (a + bx)\cosh x + (c + dx)\sinh x.$$

T 7.6   The auxiliary equation of the homogeneous equation is given by $\lambda^3 - 3\lambda^2 + 2\lambda = \lambda(\lambda - 1)(\lambda - 2) = 0$ with roots $\lambda_{1,2,3} = 0, 1, 2$. Thus the complementary function is

$$y_c(x) = A + Be^x + Ce^{2x}.$$

For the particular integral, we choose

$$y_p(x) = x(ax + b) + cxe^x = ax^2 + bx + cxe^x.$$

Substitution into the differential equation and comparison of coefficients yields $4a = 1$, $-6a + 2b = 0$ and $-c = 1$, so that $a = 1/4$, $b = 3/4$ and $c = -1$. The general solution is therefore

$$y(x) = A + Be^x + Ce^{2x} + \frac{x(x+3)}{4} - x\,e^x.$$

The three initial conditions give

$$A + B + C = 1$$
$$B + 2C + 3/4 - 1 = -1/4$$
$$B + 4C + 1/2 - 2 = -3/2$$

with solution $A = 1$, $B = C = 0$. The particular solution is therefore

$$y(x) = 1 + \frac{x(x+3)}{4} - x\,e^x.$$

T 7.7 The auxiliary equation is $\lambda^3 + \lambda = 0$ with roots $\lambda_{1,2,3} = 0, i, -i$. Therefore, a fundamental set of solutions of the homogeneous equation is

$$y_1(x) = 1, \qquad y_2(x) = \sin x, \qquad y_3(x) = \cos x.$$

Next set

$$y_p(x) = u_1(x) + u_2(x)\sin x + u_3(x)\cos x$$
$$\longrightarrow \quad y_p'(x) = u_1' + u_2'\sin x + u_2\cos x + u_3'\cos x - u_3\sin x.$$

The two additional conditions given in the example description reduce to

$$u_1' + u_2'\sin x + u_3'\cos x = 0,$$
$$u_2'\cos x - u_3'\sin x = 0.$$

Calculation of y_p'' and y_p''' using the last two expressions gives

$$-u_2'\sin x - u_3'\cos x = \tan x$$

upon substitution into the original differential equation. The three algebraic equations for u_1', u_2' and u_3' have solutions

$$u_1' = \tan x\,\frac{W_1}{W} = \tan x \qquad\qquad \longrightarrow \quad u_1 = -\ln(\cos x)$$

$$u_2' = \tan x\,\frac{W_2}{W} = -\sin x\tan x \quad \longrightarrow \quad u_2 = \sin x - \ln(\tan x + 1/\cos x)$$

$$u_3' = \tan x\,\frac{W_3}{W} = -\sin x \qquad\qquad \longrightarrow \quad u_3 = \cos x$$

where W, W_1, W_2, W_3 are defined by

$$W = \begin{vmatrix} 1 & \sin x & \cos x \\ 0 & \cos x & -\sin x \\ 0 & -\sin x & -\cos x \end{vmatrix} = -1, \qquad W_2 = \begin{vmatrix} 1 & 0 & \cos x \\ 0 & 0 & -\sin x \\ 0 & 1 & -\cos x \end{vmatrix} = \sin x,$$

$$W_1 = \begin{vmatrix} 0 & \sin x & \cos x \\ 0 & \cos x & -\sin x \\ 1 & -\sin x & -\cos x \end{vmatrix} = -1, \qquad W_3 = \begin{vmatrix} 1 & \sin x & 0 \\ 0 & \cos x & 0 \\ 0 & -\sin x & 1 \end{vmatrix} = \cos x.$$

Consequently,

$$y_p(x) = -\ln(\cos x) + [\sin x - \ln(\tan x + 1/\cos x)]\sin x + \cos^2 x$$

leads to the general solution

$$y(x) = A + B\sin x + C\cos x - \ln(\cos x) - \sin x \ln(\tan x + 1/\cos x).$$

Generalization of the procedure to a differential equation of order n is possible: we now have the conditions

$$y_1 u_1' + y_2 u_2' + \ldots + y_n u_n' = 0$$
$$y_1' u_1' + y_2' u_2' + \ldots + y_n' u_n' = 0$$
$$\vdots \qquad \qquad \vdots$$
$$y_1^{(n-1)} u_1' + y_2^{(n-1)} u_2' + \ldots + y_n^{(n-1)} u_n' = g(x)$$

so that

$$u_m(x) = g(x)\frac{W_m(x)}{W(x)}, \qquad m = 1, 2, \ldots, n$$

where $W(y_1, y_2, \ldots, y_n)$ is the Wronskian and W_m is obtained by replacing the mth column of W by $(0, 0, 0, \ldots, 1)^T$.

T 7.8   The general formula of the Laplace transform of the derivative of order n gives $(n = 4)$

$$L\left[y^{(4)}(t) : t \to s\right] = s^4 L\{y\} - s^3 y(0) - s^2 \dot{y}(0) - s\ddot{y}(0) - y^{(3)}(0).$$

Substitution of the initial conditions followed by a partial fraction decomposition now gives

$$L[y : t \to s] = \frac{e^{-s}}{s^4 - 1} = e^{-s}\left[\frac{1}{4}\frac{1}{s-1} - \frac{1}{4}\frac{1}{s+1} - \frac{1}{2}\frac{1}{s^2+1}\right].$$

The second shifting property $L^{-1}[e^{-as}F(s) : s \to t] = f(t-a)H(t-a)$ permits us to find

$$y(t) = \frac{1}{4}L^{-1}\left[\frac{e^{-s}}{s-1} : s \to t\right] - \frac{1}{4}L^{-1}\left[\frac{e^{-s}}{s+1} : s \to t\right]$$

$$-\frac{1}{2}L^{-1}\left[\frac{e^{-s}}{s^2+1} : s \to t\right]$$

$$= \frac{1}{4}e^{t-1}H(t-1) - \frac{1}{4}e^{-(t-1)}H(t-1) - \frac{1}{2}\sin(t-1)H(t-1)$$

$$= \frac{1}{2}H(t-1)\left[\sinh(t-1) - \sin(t-1)\right].$$

TUTORIAL EXAMPLES 8

T 8.1 .The variable y_2 is removed by first differentiating the first differential equation to $y_1'' = y_1' - 2y_2'$, substituting the second differential equation for y_2' to get $y_1'' = y_1' - 2(3y_1 - 4y_2)$ and finally using the first equation again to eliminate y_2. The result is

$$y_1'' + 3y_1' + 2y_1 = 0.$$

The auxiliary equation is $\lambda^2 + 3\lambda + 2 = 0$ with roots $\lambda_1 = -1$ and $\lambda_2 = -2$. Therefore

$$y_1(x) = Ae^{-x} + Be^{-2x}$$

and

$$y_2(x) = \frac{1}{2}(y_1 - y_1') = Ae^{-x} + \frac{3B}{2}e^{-2x}.$$

The initial conditions give $A + B = -1$ and $A + 3B/2 = 2$ such that $A = -7$ and $B = 6$ and we get

$$y_1(x) = -7e^{-x} + 6e^{-2x}, \qquad y_2(x) = -7e^{-x} + 9e^{-2x}.$$

T 8.2 From the description of the population dynamics, we can delineate a system of differential equations in the form

$$\dot{f} = ah - \mu_f f, \qquad \dot{h} = \mu_h h - bf$$

where $f(t)$ and $h(t)$ are the fox and hare populations respectively and a and b are two proportionality constants. Since $f = F$ and $h = H$ describes a static solution of this system we must have $aH - \mu_f F = 0$ and $\mu_h H - bF = 0$, which permit elimination of a and b such that

$$\dot{f} = \frac{\mu_f}{H}(Fh - Hf), \qquad \dot{h} = \frac{\mu_h}{F}(Fh - Hf).$$

Elimination of f, say, gives us

$$\ddot{h} = (\mu_h - \mu_f)\dot{h} \qquad \longrightarrow \qquad \dot{h} - (\mu_h - \mu_f)h = C.$$

The initial conditions are $h(0) = kH$ $(0 < k < 1)$ and $f(0) = F$. It follows from the differential equation that $\dot{h}(0) = -\mu_h(1 - k)H$ which establishes that $C = (k\mu_f - \mu_h)H$. It remains to solve a linear differential equation of first order for h. The particular solution is

$$h(t) = \frac{H}{\mu_h - \mu_f}\left[\mu_h - k\mu_f - (1 - k)\mu_h\, e^{(\mu_h - \mu_f)t}\right].$$

The hare population dies out when $h(t^*) = 0$, that is, where

$$e^{(\mu_h - \mu_f)t^*} = \frac{\mu_h - k\mu_f}{(1 - k)\mu_h}$$

or

$$t^* = \frac{1}{\mu_h - \mu_f}\ln\left[1 + \frac{k}{1 - k}\left(1 - \frac{\mu_f}{\mu_h}\right)\right].$$

The fox population can be calculated from

$$f(t) = -\frac{F}{\mu_h H}\left(\dot{h} - \mu_h h\right) = \frac{\mu_f F}{\mu_h H}h(t) + F\left(1 - \frac{k\mu_f}{\mu_h}\right)$$

valid for $0 \le t \le t^*$. For $t > t^*$, we have $h(t) \equiv 0$ (all hares have died) and the system of differential equations reduces to $\dot{f} = -\mu_f f$. As a consequence, the fox population will decay exponentially for $t > t^*$ according to the formula $f(t) = F^* e^{-\mu_f t}$ where $F^* = f(t^*)$.

Clearly, the mathematical modelling in this example was very simplistic. A more realistic approach would be to use

$$\dot{f} = afh - \mu_f f, \qquad \dot{h} = \mu_h h - bfh.$$

Here the products fh try to model the fact that foxes can not eat hares until they actually locate them. This system of equations is non–linear (due to the terms fh) and has no closed form solution.

T 8.3   Let v be the speed of the golf ball and θ the inclination of its trajectory to the horizontal ($v \geq 0$, $0 < \theta < \pi/2$). Then it follows that $\dot{x} = v\cos\theta$ and $\dot{y} = v\sin\theta$ since $v^2 = \dot{x}^2 + \dot{y}^2$ and $\tan\theta = \dot{y}/\dot{x}$. Thus

$$
\begin{aligned}
2v\dot{v} &= 2\dot{x}\ddot{x} + 2\dot{y}\ddot{y} = 2\dot{x}\left[-\frac{f(v)}{v}\dot{x}\right] + 2\dot{y}\left[-\frac{f(v)}{v}\dot{y} - g\right] \\
&= -2vf(v) - 2g\dot{y} = -2vf(v) - 2gv\sin\theta .
\end{aligned}
$$

On dividing both sides by $2v$, it is seen that $\dot{v} = -f(v) - g\sin\theta$. Similarly,

$$
\frac{\dot{\theta}}{\cos^2\theta} = \frac{\dot{x}\ddot{y} - \dot{y}\ddot{x}}{\dot{x}^2} = \frac{1}{\dot{x}^2}\left[\dot{x}\left(-\frac{f(v)}{v}\dot{y} - g\right) - \dot{y}\left(-\frac{f(v)}{v}\dot{x}\right)\right] = -\frac{g}{\dot{x}} .
$$

Multiplying by $\cos^2\theta$ and replacing $\dot{x}$ gives $\dot{\theta} = -(g\cos\theta)/v$. The initial value problem may now be restated as the first order system

$$
\begin{aligned}
\dot{x} &= v\cos\theta & x(0) &= 0 \\
\dot{y} &= v\sin\theta & y(0) &= 0 \\
\dot{v} &= -f(v) - g\sin\theta & v(0) &= v_0 \\
\dot{\theta} &= -\frac{g\cos\theta}{v} & \theta(0) &= \theta_0 .
\end{aligned}
$$

For the special case when $f = kgv$, the differential equations for the speed $v(t)$ and the inclination $\theta(t)$ are now manipulated such that v is expressed as a function of θ. We find for $v(\theta)$

$$
\frac{dv}{d\theta} = \frac{\dot{v}}{\dot{\theta}} = \frac{v(kv + \sin\theta)}{\cos\theta} .
$$

This is a Bernoulli equation which can be converted into the linear equation

$$
\frac{dz}{d\theta} + z\tan\theta = -\frac{k}{\cos\theta}
$$

using the substitution $z = v^{-1}$. The integrating factor is $1/\cos\theta$ and thus

$$
\frac{z}{\cos\theta} = -k\tan\theta + C \qquad \longrightarrow \qquad z = -k\sin\theta + C\cos\theta .
$$

The original equation for $\dot{\theta}$ now becomes

$$
\dot{\theta} = -gz\cos\theta = g(k\sin\theta - C\cos\theta)\cos\theta .
$$

From the initial condition for $\dot{\theta}$ we have

$$
\dot{\theta}(0) = -\frac{g\cos\theta_0}{v_0} \qquad \longrightarrow \qquad C = k\tan\theta_0 + \frac{1}{v_0\cos\theta_0} .
$$

Next we re–express the differential equation for $\dot\theta$ as

$$kg\,\frac{dt}{d\theta} = \frac{k}{\cos^2\theta\,(k\tan\theta - C)} \longrightarrow kgt = \ln(k\tan\theta - C) + A$$

$$\longrightarrow \tan\theta = \frac{C}{k} + De^{kgt}\,.$$

The integration constant D follows from $\theta(0) = \theta_0$ so that we finally deduce that the inclination $\theta(t)$ of the trajectory at time t is

$$\tan\theta = \tan\theta_0 + \frac{1}{kv_0\cos\theta_0}\left(1 - e^{kgt}\right)\,.$$

We note that $\theta \to -\pi/2$ as $t \to \infty$ although in practice, the golf ball will hit the ground after a finite time.

When, next, air resistance is constant, then $f = kg$ and the speed $v(t)$ and inclination $\theta(t)$ satisfy the differential equations

$$\dot v = -kg - g\sin\theta\,, \qquad \dot\theta = -\frac{g\cos\theta}{v}\,.$$

As before, we express v as a function of θ and obtain

$$\frac{dv}{d\theta} = \frac{\dot v}{\dot\theta} = \frac{v(k + \sin\theta)}{\cos\theta}\,.$$

This equation is separable and has solution

$$\ln v = -\ln\left(\cos\theta\right) + k\int \frac{d\theta}{\cos\theta}\,.$$

By expressing $1/\cos\theta$ as $\cos\theta/(1 - \sin^2\theta)$ and then using partial fractions on the latter, it is established that

$$\ln\left(v\cos\theta\right) = \frac{k}{2}\int\left(\frac{\cos\theta}{1 - \sin\theta} + \frac{\cos\theta}{1 + \sin\theta}\right)d\theta = \frac{k}{2}\ln\left(\frac{1 + \sin\theta}{1 - \sin\theta}\right) + C\,.$$

Let $v = v^*$ be the speed of the golf ball at the highest point of its trajectory (i.e., when $\theta = 0$), then

$$v = v^*\,\frac{1}{\cos\theta}\left(\frac{1 + \sin\theta}{1 - \sin\theta}\right)^{k/2} \qquad\longrightarrow\qquad v = v^*\,\frac{(\cos\theta)^{k-1}}{(1 - \sin\theta)^k}\,.$$

TUTORIAL EXAMPLES 9

T 9.1   The standard solution for this type of differential equation is
$y(x) = A \cos \sqrt{\lambda} x + B \sin \sqrt{\lambda} x$. The boundary conditions require $A = 0$
and $-A\sqrt{\lambda} \sin \sqrt{\lambda} + B\sqrt{\lambda} \cos \sqrt{\lambda} = 0$ or

$$A = 0, \qquad B\sqrt{\lambda} \cos \sqrt{\lambda} = 0.$$

The choices $B = 0$ or $\lambda = 0$ produce the trivial solution. Therefore we must
have

$$\cos \sqrt{\lambda} = 0 \longrightarrow \sqrt{\lambda_n} = (2n-1)\frac{\pi}{2} \longrightarrow \lambda_n = (2n-1)^2 \frac{\pi^2}{4}, \quad n = 1, 2, 3, \ldots .$$

The eigenfunctions are

$$\phi_n(x) = k_n \sin \frac{(2n-1)\pi x}{2}.$$

They can be normalized according to

$$\int_0^1 \phi_n^2(x)\, dx = k_n^2 \int_0^1 \sin^2 \frac{(2n-1)\pi x}{2}\, dx = \frac{k_n^2}{2} = 1.$$

Choosing $k_n = \sqrt{2}$, we therefore get $\phi_n(x) = \sqrt{2}\sin(n - 1/2)\pi x$ as the
normalized eigenfunctions.

T 9.2   As previously, $y(x) = A \cos \sqrt{\lambda} x + B \sin \sqrt{\lambda} x$. Now the boundary
conditions give

$$y'(0) = 0 \longrightarrow B\sqrt{\lambda} = 0 \longrightarrow B = 0 \text{ or } \sqrt{\lambda} = 0,$$
$$y'(1) = 0 \longrightarrow -A\sqrt{\lambda} \sin \sqrt{\lambda} + B\sqrt{\lambda} \cos \sqrt{\lambda} = 0.$$

$\underline{\lambda = 0}$. In this case the differential equation reduces to $y'' = 0$ with general
solution $y = ax + b$. The boundary conditions are indeed fulfilled for $a = 0$
and arbitrary b. Therefore, $\phi_0(x) = 1$ is a normalized eigenfunction for the
eigenvalue $\lambda = 0$.
$\underline{B = 0}$.   We obtain a non–trivial solution if $\sin \sqrt{\lambda} = 0$ which gives the
eigenvalues $\sqrt{\lambda_n} = n\pi$ and the associated eigenfunctions $\phi_n(x) = k_n \cos n\pi x$,
$n = 1, 2, \ldots$. Normalization is achieved with $k_n = \sqrt{2}$.

To summarize, the eigenvalues are

$$\lambda_n = \begin{cases} 0 & n = 0 \\ n^2\pi^2 & n = 1, 2, 3, \ldots \end{cases}$$

and the normalized eigenfunctions are

$$\phi_n(x) = \begin{cases} 1 & n = 0 \\ \sqrt{2}\cos n\pi x & n = 1, 2, 3, \dots \end{cases}$$

T 9.3 For $\lambda = 0$ the solution is simply $y(x) = ax + b$. Applying the boundary conditions gives $a = 0$ and $2a + b = 0$, with the trivial solution. Therefore, $\lambda = 0$ is not an eigenvalue. Assuming $\lambda \neq 0$, the solution is again $y(x) = A\cos\sqrt{\lambda}x + B\sin\sqrt{\lambda}x$. The first boundary condition reduces to $B = 0$, whereas the second boundary condition subsequently requires $A(\cos\sqrt{\lambda} - \sqrt{\lambda}\sin\sqrt{\lambda}) = 0$. Therefore, in order to get a non–trivial solution, we must have

$$\sqrt{\lambda} = \cot\sqrt{\lambda}.$$

This is a transcendental equation. Approximate solutions can be obtained graphically from Figure B.4:

$$\sqrt{\lambda_1} \approx 0.86 \qquad \longrightarrow \qquad \lambda_1 \approx 0.74$$

is the eigenvalue with smallest value. From the figure we can also estimate

$$\sqrt{\lambda_n} \approx (n-1)\pi \qquad \longrightarrow \qquad \lambda_n \approx (n-1)^2\pi^2, \quad n \gg 1.$$

The eigenfunctions are $\phi_n(x) = k_n \cos\sqrt{\lambda_n}x$. The normalization constant follows from

$$\int_0^1 \phi_n^2(x)\,dx = \frac{k_n^2}{2}\left(1 + \frac{\sin\sqrt{\lambda_n}\cos\sqrt{\lambda_n}}{\sqrt{\lambda_n}}\right) = \frac{k_n^2}{2}\left(1 + \sin^2\sqrt{\lambda_n}\right) = 1$$

(where we used the eigenvalue condition $\sqrt{\lambda_n} = \cot\sqrt{\lambda_n}$ in the evaluation). Consequently $k_n^2 = 2/(1 + \sin^2\sqrt{\lambda_n})$ and the normalized eigenfunctions are

$$\phi_n(x) = \frac{\sqrt{2}}{\sqrt{1 + \sin^2\sqrt{\lambda_n}}}\cos\sqrt{\lambda_n}x, \quad n = 1, 2, 3, \dots .$$

T 9.4 The auxiliary equation corresponding to a solution $y = a\,e^{\alpha x}$ is $\alpha^2 + \alpha(\lambda + 1) + \lambda = 0$ with roots $\alpha_1 = -1$, $\alpha_2 = -\lambda$. Therefore

$$y(x) = Ae^{-x} + Be^{-\lambda x}$$

is the general solution. Applying the boundary conditions provides

$$y'(0) = 0 \qquad \longrightarrow \qquad A = -\lambda B,$$
$$y(1) = 0 \qquad \longrightarrow \qquad B\left(-\lambda e^{-1} + e^{-\lambda}\right) = 0.$$

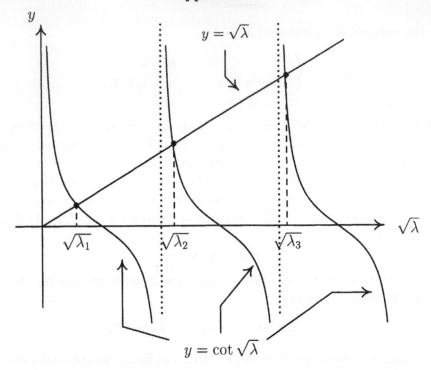

Figure B.4: Graphical solution of $\sqrt{\lambda} = \cot\sqrt{\lambda}$ from Tutorial example T 9.3 with the first three roots, $\sqrt{\lambda_1}$, $\sqrt{\lambda_2}$, $\sqrt{\lambda_3}$ indicated.

If $B = 0$, only the trivial solution emerges. If $B \neq 0$, then $\lambda = e^{1-\lambda}$ from the second boundary condition. It is easy to see from Figure B.5 that this transcendental equation has only one real–valued solution at $\lambda = 1$. The first boundary condition now gives $A = -B$. Upon substitution into the general solution, only the trivial solution $y \equiv 0$ emerges, although a non–trivial eigenvalue had been obtained.

T 9.5   In expanded form, the equation is $x^2 y'' + xy' + \lambda x = 0$ which is Euler's differential equation. We use $y = x^\alpha$ and obtain

$$\left(\alpha^2 + \lambda\right) x^\alpha = 0 \qquad \longrightarrow \qquad \alpha^2 + \lambda = 0 \,.$$

$\underline{\lambda = 0}$. The differential equation reduces to $(xy')' = 0$ which can be integrated to yield $xy' = C$ and $y = C\ln x + D$. Both boundary conditions are fulfilled for an arbitrary value of D as long as $C = 0$. Therefore, $\lambda_0 = 0$ is an eigenvalue with the normalized eigenfunction $\phi_0(x) = \left(e^{2\pi} - 1\right)^{-1/2}$.
$\underline{\lambda < 0}$. With $\mu = -\lambda$ ($\mu > 0$) we get $\alpha_{1,2} = \pm\sqrt{\mu}$ and thus the general solution is $y(x) = Ax^{\sqrt{\mu}} + Bx^{-\sqrt{\mu}}$. The boundary conditions give

$$y'(1) = 1 \qquad \longrightarrow \qquad (A - B)\sqrt{\mu} = 0 \,,$$

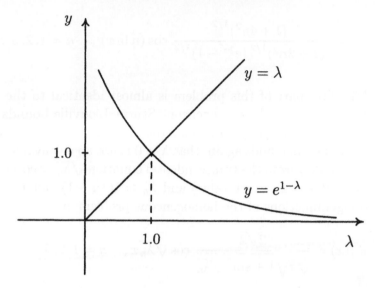

Figure B.5: Graphical solution of $\lambda = e^{1-\lambda}$ with the single root $\lambda = 1$ from Tutorial example T 9.4.

$$y'(e^{2\pi}) = 0 \qquad \longrightarrow \qquad A\sqrt{\mu}\left[e^{2\pi(\sqrt{\mu}-1)} - e^{2\pi(-\sqrt{\mu}-1)}\right].$$

A non–trivial solution requires $A \neq 0$. Consequently, $e^{2\pi\sqrt{\mu}} = e^{-2\pi\sqrt{\mu}}$ which has the single root $\mu = 0$, so only the trivial solution emerges.

<u>$\lambda > 0$</u>. Here, $\alpha_{1,2} = \pm i\sqrt{\lambda}$ and the general solution is

$$y(x) = Ax^{i\sqrt{\lambda}} + Bx^{-i\sqrt{\lambda}}.$$

The two boundary conditions now give $A = B$ and $e^{2\pi i\sqrt{\lambda}} = e^{-2\pi i\sqrt{\lambda}}$ the latter having the roots $\sqrt{\lambda_n} = n$, $(n = 1, 2, 3, \ldots)$, in view of the identity $e^{2\pi i n} = 1$. The eigenvalues and eigenfunctions are therefore given by

$$\lambda_n = n^2, \qquad \phi_n(x) = k_n\left(x^{in} + x^{-in}\right), \qquad n = 1, 2, 3, \ldots.$$

Normalization requires

$$\int_1^{e^{2\pi}} \phi_n^2(x)\, dx = k_n^2 \left[\frac{x^{2in+1}}{2in+1} + 2x + \frac{x^{-2in+1}}{-2in+1}\right]_1^{e^{2\pi}}$$

$$= k_n^2 \frac{4\left(1+2n^2\right)\left(e^{2\pi}-1\right)}{1+4n^2} = 1$$

so that finally we have the normalized eigenfunctions

$$\phi_n(x) = \frac{\left(1+4n^2\right)^{1/2}}{2\left(1+2n^2\right)^{1/2}\left(e^{2\pi}-1\right)^{1/2}}\left(x^{in} + x^{-in}\right)$$

$$= \frac{(1+4n^2)^{1/2}}{(1+2n^2)^{1/2}(e^{2\pi}-1)^{1/2}} \cos(n\ln x), \quad n = 1,2,3,\ldots.$$

T 9.6   The first part of this problem is almost identical to the task set in Example 9.3. Indeed, the homogeneous Sturm–Liouville boundary value problem can be solved by simply rescaling the independent variable in Example 9.3. In any case, one finds again, that $\lambda = 0$ is not an eigenvalue, that the eigenvalues λ_n follow from the transcendental equation $\sqrt{\lambda_n} = \cot(\pi\sqrt{\lambda_n}/4)$, giving $\lambda_1 = 1$ (this is an exact value) and $\lambda_n \approx 16(n-1)^2$ for $n \gg 1$. The normalized eigenfunctions of the homogeneous problem are

$$\phi_n(x) = \frac{2\sqrt{2}}{\sqrt{\pi}\sqrt{1+\sin^2\sqrt{\lambda_n}}} \cos\sqrt{\lambda_n}x, \quad n = 1,2,3,\ldots.$$

The expansion of the solution $y(x)$ of the inhomogeneous problem in terms of the normalized eigenfunctions $\phi_n(x)$ is given by

$$y(x) = \sum_{n=1}^{\infty} \frac{c_n}{9-\lambda_n}\phi_n(x)$$

where

$$c_n = \int_0^{\pi/4} \cos x\, \phi_n(x)\, dx = k_n \int_0^{\pi/4} \cos x \cos\sqrt{\lambda_n}\, dx.$$

Since $\lambda_1 = 1$, this case is treated separately. We obtain $c_1 = \sqrt{\pi+2}/2\sqrt{2}$, whereas $c_m = 0$, for $m > 1$. Therefore

$$y(x) = \frac{\cos x}{8}.$$

The infinite sum in the expansion of $y(x)$ has collapsed to a single term only. The reason for this behaviour is that the inhomogeneous term $\cos x$ on the right–hand side of the differential equation is an eigenfunction of the homogeneous Sturm–Liouville eigenvalue problem.

Before deriving a solution directly, let us first confirm that a unique solution for the inhomogeneous problem exists. To that end, we recognize that $y_1(x) = \cos 3x$ and $y_2(x) = \sin 3x$ as a fundamental set of solutions of the homogeneous equation. By (9.26) in Section 9.3, the existence of a unique solution for the inhomogeneous problem is guaranteed by virtue of

$$\begin{vmatrix} \hat{R}_1 y_1 & \hat{R}_1 y_2 \\ \hat{R}_2 y_1 & \hat{R}_2 y_2 \end{vmatrix} = 6\sqrt{2} \neq 0.$$

The direct solution method starts from the complementary function

$$y_c(x) = A \cos 3x + B \sin 3x .$$

A particular integral can be obtained with the choice

$$y_p(x) = \alpha \cos x + \beta \sin x$$

which leads to $\alpha = 1/8$ and $\beta = 0$. Straightforward implementation of the boundary conditions provides again $y(x) = (1/8) \cos x$.

TUTORIAL EXAMPLES 10

T 10.1   For the given functional, the Euler equation leads to

$$\frac{d}{dx}(2y') - 4x = 0 \quad \longrightarrow \quad y'' = 2x \quad \longrightarrow \quad y' = x^2 + A$$

with the general solution

$$y(x) = \frac{x^3}{3} + Ax + B .$$

The boundary conditions require $B = 0$ and $A = 1$ so that the extremal is

$$y_0(x) = \frac{x^3}{3} + x .$$

We can determine its nature by calculating

$$
\begin{aligned}
I(y_0 + h) &- I(y_0) \\
&= \int_0^3 \left[(y_0' + h')^2 + 4x(y_0 + h) - y_0'^2 - 4xy_0 \right] dx \\
&= \int_0^3 h'^2 \, dx + 2 \int_0^3 \left(2xh + x^2 h' + h' \right) dx \\
&= \int_0^3 h'^2 \, dx + 2 \left[x^2 h + h \right]_0^3 = \int_0^3 h'^2 \, dx \geq 0
\end{aligned}
$$

for some function $h(x)$ which fulfils $h(0) = h(3) = 0$. This shows that the extremal $y_0(x)$ is a minimum. The extremum is $I(y_0) = 852/5$.

T 10.2   The Euler equation is

$$2y - 4x - \frac{d}{dx}(2y') = 0 \quad \longrightarrow \quad y'' - y = -2x .$$

Hence,

$$\int_0^a \left(hh'' + \lambda^2 h^2 \right) dx = \left[hh' \right]_0^a - \int_0^a h'^2 \, dx + \lambda^2 \int_0^a h^2 \, dx$$

therefore

$$- \int_0^a h'^2 \, dx + \lambda^2 \int_0^a h^2 \, dx = - \int_0^a f^2(t) \, dt \le 0$$

and thus,

$$\int_0^a h'^2 \, dx \ge \lambda^2 \int_0^a h^2 \, dx$$

holds for $0 < \lambda < \pi/a$. Upon setting $a = \pi/4$, it then follows that

$$I(y_0 + h) - I(y_0) = \int_0^{\pi/4} \left(h^2 - h'^2 \right) dx \le 0$$

and the extremum is a maximum.

T 10.5   The first integral of the Euler equation is

$$f - y' f_{y'} = \frac{1 + y^2}{y'^2} + 2 \frac{1 + y^2}{y'^2} = A \, .$$

Consequently, $y' = C\sqrt{1 + y^2}$. which can be separated to give

$$\frac{dy}{\sqrt{1 + y^2}} = C dx \quad \longrightarrow \quad \sinh^{-1} y = Cx + D \quad \longrightarrow \quad y = \sinh(Cx + D) \, .$$

The two boundary conditions provide $D = 0$ and $\sinh C = a$ so that the extremal is

$$y_0(x) = \sinh \left(x \sinh^{-1} a \right) \, .$$

A lengthy, but straightforward calculation shows that

$$I(y_0 + h) - I(y_0)$$
$$= \frac{3}{C^2} \int_0^1 \frac{1}{\cosh^4 Cx} \left[\frac{\cosh^2 Cx}{C^2} h'^2 + h^2 \left(1 - \sinh^2 Cx \right) \right] dx \, .$$

When $0 < x < 1$ then

$$1 - \sinh^2 Cx > 1 - \sinh^2 C = 1 - a^2 \, .$$

Therefore, if $|a| < 1$ then $1 - \sinh^2 Cx > 0$ so that the integrand becomes positive for functions $h(x)$ and the extremum is a minimum.

Index

air resistance, 17, 25, 95, 103, 145
arc length, 19, 124
auxiliary equation, 30, 83
    multiplicity of roots, 39, 83
    roots of, 30

Bernoulli equation, 8
boundary value problem, 105, 107
    first kind, 107
    periodic type, 107
    second kind, 107
    Sturm type, 109
    third kind, 107
Brachistochrone, 123, 131

catenary, 125, 133
Cayley–Hamilton theorem, 98
commutation of matrices, 100
comparison of coefficients, 39
constant coefficients, 30, 83, 92
constrained extremum, 135
curves of pursuit, 19, 25, 26
cycloid, 124, 132

differential equation
    first order, 1
    order n, 81
    second order, 27
Dirac delta function, 73
distribution, 73

eigenfunction, 106
    orthogonality, 112
    orthonormality, 113
eigenfunction expansion, 116

eigenvalue, 97, 106, 112
eigenvector, 97
elastic string, 104
electrical circuit, 21, 26, 28, 69, 78, 80
electricity
    charge, 21, 28
    current, 21, 28
    Kirchhoff's Law, 21, 28
    voltage, 21, 28
equilibrium, 152
Euler equation, 128, 135, 141
Euler's differential equation, 48
exact equation, 4, 49, 86
extremal, 128
extremum, 128
    constrained, 135
    unconstrained, 126

first integral, 130
first order differential equation, 1
    Bernoulli, 8
    exact, 4
    general solution, 1
    homogeneous, 3
    initial condition, 1
    initial value problem, 1
    integrating factor, 6
    linear, 6
    particular solution, 1
    Riccati, 8
    separable, 2
    singular solution, 9
    trivial solution, 1